GAIA WEEPS

Other Books by Kevin E. Ready:

—

A New Chance (2020)

Viral (2020)

All the Angels Were Jewish (2016)

Gaia Weeps - The Crisis of Global Warming (1998)

The Big One (1997)

The Holy Koran - Modern English Translation (editor) (2014)

Credit Sense: How to Borrow Money and Manage Debt (1989)

—

and with **Cap Parlier:**

TWA 800 - Accident or Incident? (1998)

—

and writing as Sarah Sarnoff

The Disambiguation of Susan (2014)

GAIA WEEPS

by
Kevin E. Ready

SAINT GAUDENS PRESS
Phoenix, Arizona & Santa Barbara, California

See other great books available from Saint Gaudens Press
http://www.SaintGaudensPress.com

Saint Gaudens Press
Post Office Box 405
Solvang, CA 93464-0405

Saint Gaudens, Saint Gaudens Press
and the Winged Liberty colophon
are trademarks of Saint Gaudens Press

Print edition ISBN: 978-0-943039-00-8
eBook ISBN: 978-0-943039-09-1
Library of Congress Catalog Number: 9886376

Printed in the United States of America

10 8 6 4 2 3 5 7 9 1

Dedication

This book is dedicated to James Lockwood,
Lynn Margulis, Stephen Schneider, Al Gore, Jr.
and the many others who have tried to warn us.

Note from the Author on the 2021 Edition:

This story was written and originally published in the 1990's. In the twenty-plus years since the original print edition of this book was published, there have been many changes in the landscape of the argument about global warming. Unfortunately, the naysayers who refuse to see the global warming facts in the world around us still exist, and have seemingly been encouraged that nothing like the environmental holocaust spoken of in this book has actually occurred, although the evidence is growing. At least in 2021, people are no longer laughing when scientists hint that increasingly strong and frequent tropical storms and rampant wild fires may be caused by global warming. However, the basic facts and scientific basis of this novel remain, and have been solidified by decades of further research and evidence. The prescience and purpose of the original story remain as well.

For the eBook edition in 2013 and for this 2021 print edition it was decided not to update the scientific discussions of the original book, as it is believed the passage of time has only strengthened and confirmed their voice. However, we have done a little story line enhancement and plot update. Things like replacing the author's old duty station, the now decommissioned carrier, USS Kitty Hawk, with the USS Ronald Reagan, not having Al Gore as the Vice-President, a minor character having the same name as a now-famous actress and various other story and timeline upgrades to enhance the reader's enjoyment by removing story and plot "oopses" which the passage of twenty years have engendered. However, it is amazing how much of the story remains vital and relevant two decades hence.

Along those same lines, the wisdom of hindsight allows us to replace 2012's Hurricane Sandy that devastated much of the American East Coast with the Hurricane Clark that was discussed in the 1998 edition of this novel. No apologies for the eerie literary prediction about Sandy, but my hopes and thoughts do go out to Sandy's victims, as well as to the more recent victims of Hurricane Ida. Tropical storms do, indeed, seem to be proliferating. As I said in the afterward to the 1998 edition (see it at the end of the book), I can only pray that the other bad things mentioned in my global warming story are short on prophecy, but I fear they are not.

Thank you for reading my book. Please think about what it says.

Kevin E. Ready

March - Boston, Massachusetts

"Boston Approach, AtlanticAir 731 Heavy passing six point five for four," radioed First Officer Jim Thurman as he watched the altimeter passed 6,500 feet and reset the altitude alert bug to 4,000 feet.

"AtlanticAir 731 Heavy, Boston. Radar contact. You are cleared for the ILS Two Seven approach. Intercept the glideslope prior to KLANE. Winds three oh five at ten gusting to fifteen, in heavey fog and light rain. RVR sixteen and deteriorating. Altimeter two eight point four five."

"Two eight point four five. Cleared ILS Two Seven, for 731 Heavy" responded Thurman. He turned to Captain Bruce Mathieson. "Looks like this is going to be a blind one."

AtlanticAir Captain Bruce Mathieson, a veteran pilot with 30 years of experience, nodded his head. Both pilots reset their altimeters to the lower pressure settings. They continued stepping through the landing checklist. Mathieson was concerned, but not troubled, by the worsening weather.

They had been notified that La Guardia and Kennedy were both socked in, below the weather minimums, with the enormous approaching storm system from the South as soon as he had checked in with Gandor Center after the crossing of the Atlantic from London's Heathrow Airport. Luckily, there were only a few passengers on the big, four engine, Airbus 340 to be upset by the change in destination. This flight was basically was a nighttime shuttle to return the plane to New York for tomorrow's daytime revenue producing mid-morning run.

AtlanticAir did not fly any international flights into Boston, and Mathieson flew exclusively on the Airbus 340's to Europe. So, he was not too familiar with the approach, but he had already checked the various Jeppeson charts which now sat on both control columns and his co-pilot's knee. He looked at his watch, flicking the button to switch the digital readout from London time, it was 4:05 A.M. Eastern time.

Mathieson was weary from the long flight, but the approach went smoothly. Boston's Logan Airport had very little traffic in the early morning hours. The approach controller, a young woman with a cheerful voice, would have just come on shift at 4 o'clock and had very little other traffic to worry about.

The flight deck crew completed the landing checklist.. They continued, as always, to cross-check their instruments and the approach plate diagram to ensure the large airplane was precisely positioned as it descended toward the runway.

"Boston, 731 Heavy, we're over the Outer Marker," Thurman broadcast.

"731 Heavy, roger over RIPIT, cleared to land Runway Two Seven."

"Cleared to land for 731 Heavy."

Bruce Mathieson knew they should have been able to see lights of downtown Boston to his left, but saw nothing. The weather was getting thick indeed. Logan Airport had its runway approach lights and strobes on pylons in Boston Harbor and the surrounding wetlands. The approach brought the plane in over the harbor. The green symbols on the computerized, heads-up display showed the runway and flight path markings although there was only the cloud shrouded reflection of the airplane's landing lights beyond the windshield.

The crew watched the tenths of miles to touchdown roll off on their displays. They still had not seen any lights. Decision height was now less than a mile away. The rules required them to have positive identification of the approach and runway lights before they were procedurally allowed to pass the published decision height of 200 feet. This was the ultimate in nerve-wracking duties -- flying through clouds at night without any sight of any lights, knowing the ground was a mere few hundred feet below them, and trusting the computer chips in the airplane's black boxes were correct.

At two hundred feet, the specified decision height, Mathieson caught the characteristic pattern of the bright white approach lights and the sequencing strobe lights leading to the runway. The lights were especially dim and flickered though, for no apparent reason, like they were obstructed by something. Was the ground fog that thick?

The parallel track of runway lights disappeared into the fog ahead of them. Why did those damn things continue to flicker, Mathieson asked himself. They looked correct except for that flicker. He pulled back on the small, side stick. The nose of the large transport rose to the command as he retarded the throttles to idle in their flare for landing. He dropped the big plane onto where the runway had to be. This was the true gutwrencher of a night, instrument landing.

He was settling the plane onto the runway when he saw the shimmer of the water around the runway lights. Both pilots saw the danger at the same time.

"It's water. Pull up!" Thurman shouted.

The co-pilot reached for power at the same time Mathieson pulled back on the arm rest joystick to pull the Airbus 340 up.

Too late. The large airliners were never known for their agility, but nothing would have prevented this.

The wheels of the main landing gear dove into the 14 inches of water covering the runway. The drag of the water on the main gear pulled the plane's nose down sharply. The added engine thrust of their aborted landing attempt and their pitch up at the last few seconds only increased the impact with the submerged runway. The nose gear plunged into the water and hit the runway hard.

The passengers were thrown forward in their seatbelts. Mathiesen's weight strained the shoulder harness. He struggled to reach out to close the throttles and shutdown the engines.

Under the combined pressure from the water and the odd angle of impact with the runway, the nose gear collapsed. The inboard engines, slightly lowered on the wings than the outboards, sucked up the water beneath them. The large fan blades, designed to move large volumes of air and be tolerant of heavy rain plus bird impacts, were not designed to move water. The blades sheered off like balsa posts hit by a bat. The simple but effective containment shroud kept the fan blade remnants from spraying into the fuselage. The flash of fire was instantly extinguished by the mass of water under the airplane.

The now disabled airliner was just a large mass of aluminum with 43 humans inside. It careened, twisted, gyrated down Runway Two Seven. Captain Mathieson was able to shutdown the outboard engines before they became involved when the main landing gear both sheared off under the incalculable loads applied by the water and flailing on the runway.

The plane stopped after only a few hundred yards. The underbelly of the nose was ripped and torn from its slide across the flooded concrete, but the very water which caused the crash now cushioned the plane from the sparks which could have created an inferno. The Airbus was sideways with its nose off the runway in the deeper pool of water beside the runway.

Fuel spilled from engine pods and wing tank, but the fuel floated on the lapping waters of Boston Harbor.

Near the far end of the main runway, they were still not within sight of the tower due to the fog and rain. The tower obviously had not seen the two or three feet of water on the lower end of Runway Two Seven.

The tower controllers, watching their ground traffic control radar as they always did with any aircraft movement in bad weather, saw the short stopping distance of AtlanticAir 731. Two quick radio calls that were not answered confirming that something was dreadfully wrong although they could not see it. Two of the four tower controllers hit the large red crash button on their consoles.

The airport fire and rescue vehicles responded quickly. They, too, hit the water about a half mile out on the taxiway and only two of the biggest vehicles, the massive LaFrance fire trucks, made it out to the plane without shorting out engine wiring or flooding the engines in the waist deep water on the taxiways and runway nearest AtlanticAir 731.

Those passengers who had ignored the seat belt signs had been thrown about by the impact. Luckily, a broken collar bone was the only major injury.

The passengers of AtlanticAir 731 had to disembark via inflatable slides into the cold waters of the newly larger Boston Harbor, all would remember the unexpected water-park ride down the slide from the darkened plane into the saltwater.

By sunrise, the waters had receded with the ebb tide, but the full brunt of the late winter storm had hit, making it difficult for the National Transportation Safety Board (NTSB) to begin its investigation.

It was a unique air crash investigation. There had been previous incidents of planes landing short or long of a runway and ending up in the water, but never one in which plane had crashed into relatively deep water on the runway.

The NTSB finally determined that the crash was a unique circumstance of the unusual and unprecedented high tides combined with the unexpectedly strong storm system and its uniquely low pressure and abnormally high storm surge pushed ahead of the system. Some fault was found with the ground operations at Logan International for not having anyone on the ground check the runway before the crash. But, as any sailor or surfer knows, high tide comes in very quickly and tidal records from the Coast Guard showed that the high tide had only been up for fifteen or twenty minutes before the crash.

Logan International Airport hired a new civil servant whose job it was to circle the airport's ocean perimeter once every half hour during hours of darkness to check on tidal flooding. Hopefully, his services would be an unnecessary precaution.

Although it was not included in the final report of the NTSB, an internal memo by NTSB staffers noted that Logan International was only one of perhaps twenty major metropolitan airports in the U.S built directly on tidelands or near sea-level land. The landing threshold for Logan's Runway Two Seven is a mere 13 feet above mean sea level. Another item in the internal memo, set for later review, was the fact that no one had been able to document a reason for the unprecedented high tide that had been the real cause of the accident.

March - Enroute Washington, D,C, to Denver, Colorado

Andy Knowles knew it was silly that he still had subtle pangs of conscience every time he took a seat in the first class section of an airliner. His personal philosophy and his modest upbringing made him more conscious of the price difference between an economy class and first class ticket than most of his peers were, but the notoriety of his abortive run for the presidency a few years before had left him with little choice.

Senator Andrew Knowles had tried to take a tourist class flight out to Seattle the year before, but he had spent the whole flight fending off autograph hounds and unwanted political advice from the other passengers, rather than reading the speech he had been trying to memorize for the trip. With a few exceptions the first class passengers gave him his privacy and the flimsy curtain back to the rear compartment somehow kept the teaming masses back. Besides, only the first-class seats gave his long legs a place to rest.

On this trip he had nothing to study except one file in the leather portfolio. After accepting a his breakfast tray from the stewardess who was overly courteous to him, Andy set to work on the file on a CD in his laptop. He had not looked at the file in the year and a half since he had used its information to push through a supplemental amendment to the National Oceanic and Atmospheric Administration's (NOAA) budget. A quick glance at the file showed that his Administrative Assistant had made sure someone had updated the contents of the file for this trip with magazine clippings and NOAA fact sheets. A good "A.A." made life much easier in Washington.

Andy had at first pushed the original NOAA budget amendment because the prime contractor for the new computer it funded was located in his home state, Connecticut. The GAIA project had been proposed by the Vice-President, then a Senator and a close friend and ally of Andy in the Senate, who was fervently in favor of the project as important to the environmental program he had made his *cause celebre*.

After Andy got familiar with the project he had used the full weight of his sub-committee chairmanship to push through the interesting project for its own merits and not merely because of the economic impact on Connecticut or the influence of his friend. The two Colorado Senators and all of its members of Congress had jumped on the bandwagon along with many others interested in either the environmental significance of the project or the pork barrel impact for their home states. The fact that the University of California played a major part in the project had put the votes over the top with California's four dozen-odd members of Congress. Even a billion dollar project was easy to push through, if the dollars were spread to enough states.

In getting the amendment pushed through, Andy had worked closely with the scientist who had called his office two days before this flight. Andy's scheduling secretary had read Ken Brady's name along with several others who claimed to have "urgent" need to talk to the senator. Out of all the names on the list only a powerful congressman, a union leader and Ken Brady were called back by the Senator himself. The rest received a phone call from one of Andy's staff expressing the Senator's regrets with an explanation of his busy schedule.

"Hello Senator. Thank you for calling." Ken Brady's voice had been full of excitement as he spoke to Andy Knowles.

"It's nice to hear from you. I was a little surprised at the urgency of your message. What's the matter, didn't we appropriate enough money to keep that super computer of yours in microchips?"

"No, no problem there. We have everything we need. What I need to talk to you about is totally different and much more important."

"I'm all ears."

"Well, I've gone over what I need to tell you a hundred different ways and I haven't been able to come up with words to explain what is happening. I also tried to get hold of the Vice-President, but her staff is even tougher to get through than yours. I know it is a lot to ask, but is there any way you could come out and see for yourself. If you could see our facts, I'm sure you would appreciate the importance of what we have here."

Andy was not sure he understood the scientist, "We're in the middle of the legislative session. You want me to fly out to Boulder? What could possibly be so important that you can't tell me over the phone?"

"I know I probably sound like a nut. But I assure you I'm not crazy. We have the first full set of GAIA information and it's earthshaking in its significance. We can't get the bureaucrats at NOAA to listen to us and we have been told so much as to keep quiet about it and we don't want to go to the press with something like this, its too important. Word is that anything that might rock the boat on the issue of recovery of the domestic economy had better keep quiet and what we have discovered could sink the damn boat, if it is not handled right. I thought that a politician of your stature and understanding of GAIA would be the best person to help us and possibly get through to the Vice-President or the President."

"It's that important?"

"Senator, the reason I called you is that you know me, you know I'm not a flake or a nut case. I assure you that if you are able to make proper use of the information we have it could be one of the most important acts of your life. If you can't help us, it may mean the end of the world. I would call that important. Wouldn't you?"

That had been the day before yesterday. Andy's wife, Molly, had turned down his offer for her to come along for a day's skiing. Her schedule in court was as busy as his own schedule on the Hill. His staff had been incredulous when he had told them to schedule the flight to Colorado and pull the GAIA file. One of his pet public works projects was coming up for a vote soon and he might miss it. But his A.A., Derek Shaw, had a sense of when Andy could be talked out of something and this obviously was not such an occasion.

Senator Knowles watched out his window with interest as the plane banked into the Denver airport. An avid skier, he had flown into Denver many times. The high peaks to the west sparkled in their white and purple beauty, but Andy was surprised that the ground below him had no snow cover. He remembered the white landscape when he had landed a year ago in March. However, he also remembered his colleague in the Senate from Denver bragging about Denver's beautiful climate, three hundred days of sun per year and quick thaws after winter storms. He had commented on how non-Coloradans often confused Denver's climate with that of the mountains a hundred miles west and how they expected ski slopes in the city. The numerous heavy snowstorms the city had suffered through in recent years had not helped that image. With or without snow, it certainly did seem to be a beautiful day in Denver.

As he turned from the window, Andy noticed the stewardess who was seated facing backwards in the little jumpseat by the door had been watching him. She gave a quick, embarrassed smile and averted her eyes. Andy set about putting the laptop back into the case. He supposed that it was just as well that he was happily married, it would be very difficult for a famous senator to make a pass at the pretty stewardess in public. But then again, his colleagues in Congress somehow managed regularly enough to do a lot more than flirt. Molly's law firm alone had one or two congressional spouses as divorce clients or opponents and roving eyes were quite often the cause.

As the plane bounced its way onto the Denver runway, Andy again looked out to the west. His view of the mountains was now stained by Denver's slight red cloud of winter smog. He would soon be above the smog again, as he drove into the foothills above Denver to GAIA's windswept home.

———

March - Padma Delta, Bangladesh

Captain Akim Darpheel surveyed the outcropping of land as his boat passed it. The huddled masses of the living at first shouted and cheered when they saw his boat flying the national flag of a red orb on green approach on its way through the waters of the great river delta formed by the Ganges and Brahmaputra rivers, the combined river known as the Padma.

The people thought that certainly the government had finally come to help. Then they stared forlornly as they saw the uniformed men in the boat were not coming to them. A few of the groups of living shouted epithets and curses at Darpheel and his men when they passed them by. What did they expect him to do with six men and one boat, facing tens of thousands, if not millions, of pitiful survivors?

The corpses, both human and animal, were so thick in spots that they blocked the shores of the waterway, making it impossible to land and sometimes to navigate the boat without bumping into bloated, fetid chunks of human being or water buffalo and the other debris. But, at least the dead were silent and only tormented him when he allowed himself to look.

"The poor bastards."

Darpheel turned to the voice, that of Sergeant Quist, his half-anglo assistant. He shouted back, trying to make himself heard over the boat's engine, "I wish there was more we could do."

"I can't help but wonder if maybe the dead are not the lucky ones. These people have nothing and I doubt if the government has much they can send down here. Why, half of these survivors may die, too, in coming weeks."

"True, the dead were now beyond suffering and in the presence of Allah." Darpheel spoke the words about God without any of the intensity he should have. Feelings of grief, pity and frustration churned within him. He did not have much praise for a God who would allow this.

Darpheel could do nothing more for either the living or the dead. Occasionally he had been able to maneuver the motor whaleboat over to a small outcropping of land or a rooftop and pull a survivor from his or her clutching perch above the waters. But then he could only take them to the several larger areas of unsubmerged land in the vicinity in hopes that someone from Dhaka would be out to bring food, blankets, shelter, rescue; everything that these miserable people had nothing of.

And now he was even afraid to offer this little assistance. On his last landing to drop off rescued survivors at an old jute mill, which was significantly higher than the rest of this delta land, a group of men on the island that was now where the jute mill had been, rushed the boat in hopes that they could overpower Darpheel's crew and escape their watery prison in the boat. Seeing the men rush the boat, the rest of the survivors also raced down to where Darpheel had tried to put ashore. Only a last minute retreat and burst of gunfire had kept the boat from being swamped by an onrush of humanity, slipping and stumbling through the mud, debris and bloated bodies that littered the shore of the outcropping. He had been forced to allow the survivors he had tried to land at the jute mill to stay in the boat. He would take them back with him. If he had enough fuel to get back.

"Sergeant, cut the engine to half power, we need to conserve fuel, if we expect to get all the way to Narayanganj." Darpheel shouted above the noise of the engine. The sergeant nodded.

Neither Captain Darpheel nor his men were used to such duty, if anyone could be used to such as this. They had been with the headquarters staff of the Bangladesh Army in Dhaka only last week. But there had been an emergency response plan drawn up after big typhoon two years before had shown the country was unprepared to deal with rescue and relief operations in the face of such a disaster. This plan had called for military and other government personnel to be trained in search and rescue and to be assigned to specific areas of rescue operations. The plan had called for Captain Darpheel and his men to report to the docks outside Narayanganj to set out in the motor whale boat to do damage assessment and radio word back to Dhaka.

They had known the typhoon threatened them for days and that it was going to hit nearby for at least twenty four hours. But they had not known exactly where it would hit until about eight hours beforehand. At first the monstrous weather system had drifted west, appearing as though it might make landfall near Calcutta. Then it veered east as though it would hit Chittagong, which had borne the brunt of the '91 typhoon that had killed so many and set records for storm intensity and wind speed. Finally they saw that it would take the middle ground and head straight up the Brahmaputra River into the very center of Bangladesh.

Darpheel and his men had arrived in the Army bus along with others assigned to the rescue duties. They got to the waterfront area just hours ahead of the storm and when it came they took shelter with thousands of others in the newly constructed concrete shelters which doubled as warehouses. The bus had returned to Dhaka with children from an orphan's school aboard.

As they waited in the shelter, they listened to the national radio broadcast warnings of the storm. The radio had said that this storm appeared to be even bigger than the typhoon of '91. Winds were expected to reach 210 kilometers per hour and a tidal serge of eight meters was possible. They had been glad the shelter was on concrete pilings. As the storm approached and word of its expected ferocity spread the shelters had filled until there was no room to move in the huge warehouse. As the winds increased they heard the cries of those still outside the massive steel doors who were locked out and would have to find shelter elsewhere. The police captain in charge of the shelter had his orders and when the shelter had filled to bursting point he simply closed and locked the doors from the inside. A few of those inside complained that family members had not yet arrived, but it was clear that there was no more room for anyone. Darpheel and everyone else inside had blocked the thought of those outside from their mind.

The storm hit its full stride in early evening. The electricity to the shelter was cut off in a burst of intense wind. A howling wind that was unlike anything he had ever heard echoed through the shelter. Children screamed in the dark. Occasionally there would be another outsider screaming from beyond the doors and they heard some people on the steel paneled roof above them, but this sound was soon drowned out by the colossal wind and the reverberation of large objects and debris hitting the concrete and steel of the shelter.

The police and soldiers had battery lanterns which had occasionally cut through the dark as Akim watched. Each such beam reflected the frightened faces of a thousand souls. The lucky ones.

With the noise of the wind came occasional cracks of thunder, but even the thunder could not make itself heard well above the wind. Eventually the unmistakable sound of pounding surf added itself to the din they were bombarded with in the shelter.

The crashing of waves brought additional screams of panic to both children and adults, for the shelter was two blocks from the main pier and Narayanganj was twelve kilometers up the ship channel from the open water of the main river. For them to hear the sounds of waves crashing must mean that the tidal surge had hit.

The storm had continued throughout the night. Few of them had slept at all. Toward dawn they had heard the storm's winds and thunder abate. Captain Darpheel and his men had listened to the field radio they had brought with them throughout the night. They had heard something of the destruction being wrought, but nothing prepared them for the sight of destruction when they opened the doors at dawn.

The wind was still blowing, but in comparison it now seemed tame. Darpheel and his men had managed to push out before the bulk of the crowd and they saw what was left of the port area.

Debris was everywhere. Many unreinforced warehouses were in ruins. Trucks, cars and numerous boats from the waterfront were tossed about like discarded toys. And the bodies... .

The hundreds of bodies. At first Darpheel and his men, along with the police, had run from corpse to corpse seeing if they could help, but it soon became clear that all those who did not move were dead. Triage was simple.

Captain Darpheel had radioed the report of the destruction to Dhaka. He had had trouble getting through on the radio net. Dozens of Army patrols were trying to report the same thing to headquarters and on top of that the primary headquarters station was not on the radio net, an alternate had taken over net control. But, in spite of the destruction in Narayanganj itself, Darpheel was ordered to carry out his assigned task of proceeding into the delta and reporting damage.

The warehouse where the patrol's boat and equipment had been stored had been spared any structural damage but it took the better part of an hour to remove debris and sort out the soggy equipment that had been under water much of the night.

Captain Darpheel's patrol had been one of two Army patrols that was supposed to be supplied from the warehouse cache. The other patrol was not to be seen that morning. Darpheel had taken the opportunity to sort through the gear and take the best for his crew.

The gray motor whaleboat had the words "U.S. NAVY S/N 34297" only partially obscured by an overcoating of paint. The gas cans and emergency supplies also had American and British military markings on them. This cache was part of the world response to a previous typhoon disaster which always seemed to catch his country ill-prepared.

Darpheel had watched as Sergeant Quist and the men, with assistance of the police and some of the crowd still wandering in shock in the debris of the waterfront, moved the big boat over to the edge of the jetty and lowered it into the water below with hand winches. The activity of Darpheel's men had been the first spark of positive action anyone had seen, so many had tried to help. Darpheel had taken advantage of the opportunity and had organized a human conveyor belt to move the gas cans and provisions from the warehouse to the jetty and then down the cargo net into the boat.

When the boat was as full of supplies as Darpheel dared, he had shut the Army warehouse and turned the keys over to the police captain with instructions to wait until noon for the other Army team to arrive and if they had not shown up by then to use the stores as he saw fit for the people of Narayanganj.

Darpheel had then climbed down the net and into the boat. When they had come down for training some months before the water level had been ten feet lower on the quay wall, the water was so high he hardly needed the net to climb down. He had been sickened to see that the bodies of what appeared to be a woman and child were stuck between his boat and the jetty wall. As they had pulled away from the pier one of the men had taken up position in the bow with a pike to insure that they did not hit the floating debris or bodies.

———

March - Boulder, Colorado

The cluster of buildings sits perched on the edge of the high mesa overlooking Boulder, Colorado. The numerous satellite dishes and the angular, ultra-modern architecture confirm the scientific pursuits of the complex. The local people and the staff call the complex "N-CAR". In the rarified air, brisk winds and expansive sky of the mesa, it seems the bureaucrats and politicians responsible for its placement for once were correct in the location of the National Center for Atmospheric Research.

Pulling his Blazer into the parking space labeled "Director, GAIA," Ken Brady got the usual feeling of satisfaction at having a reserved parking space that could only come from the years of perpetual struggle to get a parking space at the university. He gathered the ever present computer printouts, laptop and his briefcase. As usual he had to square his shoulders and clutch the printouts in the stiff morning breeze that rolled off the Front Range mountains.

The few workers from the midnight shift were just leaving as Ken made his way up the walkway between buildings. He nodded greeting to the two young girls who worked the midnight shift as data entry clerks for his GAIA team as he turned up the hill to the building that housed GAIA. Avoiding the elevator, Ken intentionally took the stairs, two at a time. A man of moderate build and a receding hairline, he refused to allow himself to attain the midriff paunch that so many of his fellow scientists did after years of sedentary research tasks. At 39, he still could fit perfectly in the uniforms he wore as a young Air Force weather officer fifteen years before.

As he crossed the GAIA lobby on the third floor he passed the display of the Gaia emblem on the wall. When Ken had set up the office and needed an official seal for the government operation, his wife Ginger had been so happy that her profession as an artist had finally had some practical use in her husband's world of science. She had done a good job. The image of Gaia, the ancient Greek goddess of the earth wielding her power over the elements, the sea and the land captured the character of the work of the Geophysical and Atmospheric Interactive Analysis (GAIA) team. Ginger had been quite embarrassed when asked if the golden haired goddess was a self-portrait, but the resemblance was uncanny. Ken did not mind the resemblance.

Crossing through his team's work area he noted that the big computer was at work. The hum of the Number One tape drive and printer indicated that the GAIA staff in La Jolla was doing an early morning data dump of oceanographic data. After their lengthy phone call last night, John Colton had apparently gone into work early in California to make sure that Brady's staff had every possible piece of data for the upcoming hullabaloo Ken and John were planning to create.

"Morning, Conchita." Ken said as he passed the cubicle in which his young computer graphics specialist was already hard at work.

"Good morning." Conchita Alvarez was an atypical computer expert. Her youthful and attractive Hispanic looks and her effervescently cheerful demeanor did not jive with what one usually expected from a high tech wizard and expert computer hacker who played a key role on the GAIA team. Her tenacious work habits were typical of the team, however. Everyone loved their job and believed in the mission.

"Everything all set for our visitor?"

"Sure thing. I have all of the ice cap stuff re-done, but I'll have to wait on the ocean thermals," she stopped to motion toward the whirring computer drive across the room, "San Diego is sending an all new dump of ocean data. I'll have to wait for everyone else to process that before I can re-do the global geophysical graphics."

"I know, I talked to John Colton at home last night. Just have a good presentation with full hard copy back-up ready for us by ten. The senator will be ready for our dog and pony show then. Make sure the big monitor in the conference room is up and operating, too. You and Josh may want to do a trial run. If you have time."

"That depends on how fast the update on the new data takes. But, we'll try."

Ken nodded acknowledgment and went on to his office. His secretary was not in yet. He spread the sheaf of notes out across the table. Before sitting down he grabbed his old Air Force mug and went for coffee. He had a lot of work to do before the senator got there and he had been up late that night. A cup of coffee was needed.

———

March - Drake's Passage

The Mate of the Watch had reported the contact at some thirty miles at the extreme outside range of the navigation radar. The visual sighting was some minutes later. Now, everyone aboard the Japanese merchant vessel was on deck watching the sight.

The sun had risen when the contact was about ten miles out on the western horizon and the first rays of ruby sunlight had hit the object. Word had quickly spread through the crew to come topside.

Captain Furosawa had altered course to pass to the north. The object had a northerly drift of about two miles per hour and the original course of the ship had been a collision course. A few degrees change of course to the north and a couple of turns on the shaft had opened the closest point of approach to two miles, off the port beam.

That was were it was now. The crew and the few family members who rode the ship lined the port gunwale along with the half dozen Japanese students who had booked steerage passage from Yokohama to Sao Paulo and back. The Dekiru Maru frequently made this run with the holds full of Japanese electronics on the way down and frozen Brazilian beef or timber on the way back to Japan. The voyage had been as uneventful as always, until now.

When the iceberg crossed the beam, Captain Furosawa said to the Mate of the Watch, "Cut the engines to bare steerageway."

"Yes, Captain," the mate said as he rang the reduction in speed into the engine order telegraph.

A pride of whales had started to surface between the ship and the iceberg. The delighted spectators shouted their approval of the whales' antics.

Furosawa took the seldom used sextant off of the navigator's shelf. He had been using GPS, LORAN and SATNAV to navigate for so long the old sextant seemed clumsy in his hands. The only time they ever took it down was to check off an apprentice for his mate's qualification checklist. Now, Furosawa used it to take the angle from the water line of the berg to its peak.

When he got the angle he stepped to the radar and took a reading of the distance. With these figures he did a quick trigonometric calculation of its height. Certainly his rusty math skills were wrong. He asked the mate to check his work. The younger officer confirmed his calculation. The iceberg topped out at slightly over a thousand feet above the waters of the Drake's Passage.

The left or eastern side of the iceberg was ragged and sparkled in the light of dawn. The right side was in shadow and reflected the indigo purple of the twilight western sky, but appeared to be virtually flat. Furosawa surmised that the flat side had been the top of a flat ice floe and some thicker ice on the bottom of the berg had tilted the pinnacle of the iceberg up and out of the water.

As spectacular as this sight was, it was what he could not see that struck awe in Furosawa. With the old formula of one seventh of the iceberg above water and six sevenths below that would mean that below the water was six times the bulk of ice in the thousand foot monolith he now watched.

As he gave the order to increase speed, they started to pull away from the berg, Furosawa told one of the men to use the video camera they kept handy in case of maritime accidents and take some pictures of the mammoth. He then penciled a radiogram to Chilean Navy base at Puento Arenas, informing them of the sighting and reporting its size.

Furosawa turned the deck back over to the Mate of the Watch, but not before giving instructions for an able seaman to be stationed on the forecastle on iceberg watch until they cleared the Passage and headed northwest. In spite of the latest in navigation equipment, there still remained an ancient fear of icebergs in every sea captain's heart. And this iceberg had been truly awe-inspiring, if not fear inspiring, He would take no chances.

———

March - Padma Delta, Bangladesh

They had been in the boat nearly two days. Captain Darpheel had given out the extra food and water before the first day was out. His orders had been to check on the villages downstream on the main channel of the Padma River, as the Bangladeshis called the combined Ganges and Brahmaputra rivers, beyond Chandipur and then return to Narayanganj.

He had carried out his orders as well as possible, but the devastation of the downstream area was so total that he had trouble finding were the various villages had been and Chandipur itself was hardly recognizable. The city had been flattened and parts were still under as much as ten feet of water when he arrived the end of the first day out. Much of his extra provisions had been dispensed to survivors on rooftops in Chandipur.

Captain Darpheel had gone beyond Chandipur on the eastern shore of the main channel of the Padma. He dared not venture further in his boat. The main channel of the mighty river, even this far out in its delta was churning with floodwaters from farther north in Bangladesh and Indian Assam where the typhoon had dropped its massive cargo of rain as it spent its power over land.

They had seen every sort of debris and flotsam coming down the mighty river. A houseboat, or what had been a houseboat, was seen spinning its way out to sea. The people on the houseboat had signaled with waving arms and even a flare gun, but Darpheel knew his little boat was a poor match for mid-channel current of the Padma at flood stage. The souls on the houseboat would have to wait for whatever fate had in store for them; Darpheel could not help, except to radio their predicament to Dhaka.

Akim Darpheel had ignored the call for help and ordered his coxswain to turn about and find something sturdy to tie up to for the night. They dared not try to navigate the debris filled channels in the dark.

Akim and his men awoke at the first light of dawn. After the sleepless night before they had slept soundly in the gently rocking boat. They awoke to discover that the water levels had receded considerably and the tree they had tied up to was now almost on dry land. They pushed the boat off with an oar and got underway.

They had each eaten one of the American MRE ration kits they had saved for their own use. His men marvelled at the chemical heating pads that the Americans had put in to heat the meat entree. Darpheel wondered what the people he had given these rations to the day before had done with the chemical pouch if they did not speak English to read the instructions. The tiny pictograph instructions were worthless. He suspected that several hapless peasants would have tried to eat the warming chemicals.

Morning had brought calm weather and reduced current in the channel back up to Narayanganj. Once they had seen a large white seabird calmly floating down the channel on what at first appeared to be a little pillow. As it drifted closer they saw that the pillow was in fact the belly of a naked woman, either pregnant or already bloated, possibly both. Her face was not visible in folds of what had been her garments now covering her shoulders and face, but her breasts protruded from the water as did her rounded abdomen upon which the big bird sat. It had been nibbling the stuffing from his makeshift raft.

Even in the quagmire of emotions they had experienced with this disaster, this was more than the devoutly Moslem soldiers could take. Both Sergeant Quist and another man rammed the bolts of their weapons home. In a spray of rifle fire the body of the bird disappeared into the waters and corpse of the woman continued on its way unmolested.

Toward the end of the second day out, the boat approached a stand of poplars sticking out of open water to the south of what Darpheel assumed was the channel. The waters still covered so much land that it was a guess as to what was channel and what was flooded field, only the few standing trees and ruins gave hints.

At first it appeared that there were monkeys in the trees, but this was too far from jungles near the Burmese border where the only remaining monkeys lived. Darpheel ordered the boat over to the line of poplars.

As they looked up into the tree they saw three children, two in one tree, one in the other. They were naked and so tired they did not even cry out. One child could have been no more than a year and a half. They maneuvered the boat over between the two trees and tried to coax the children down.

The oldest child motioned to her rib cage and Darpheel saw now that she had been tied to the tree as a last attempt by their now missing parents to save them from the waters of the typhoon. One of his men scaled the tree, cut her free and carefully lowered the young girl to the boat.

As they moved to the next tree, another man climbed it. Darpheel saw the man cut the cord tying the infant to the tree and another man reached for the baby. Darpheel saw by the limp body that they were too late. However, the other child, a boy, was alive and climbed down the tree himself and into the boat.

They gave the children water and food. One of the women they had kept in the boat from the jute mill incident took charge of the children. When no one was looking, the Sergeant lowered the body of the baby wrapped in a weighted burlap bag into the water.

It was twilight of their second day when they finally rounded the last bend and saw the docks of Narayanganj. A large patrol boat flying the flag of Bangladesh was tied up at the quay wall they had launched from. The patrol boat was battered and its radar mast missing. It had come down from Dhaka while he was out in the delta. It was loading supplies.

Some of those people doing the loading appeared to before foreign soldiers, Europeans. He saw police and army personnel at either end of the quay, forcing back mobs of hungry, forlorn people eager to get at the supplies being loaded on the boat.

The coxswain pulled the whaleboat in behind the patrol boat. A Bangladesh Navy seaman on the quay ran to grab the bow line and another the stern. Captain Darpheel radioed his final report to Dhaka and climbed out of the boat. A Navy petty officer on the stern of the patrol boat saluted Darpheel and he returned the salute.

As a last thought before he left the quay, Captain Akim Darpheel looked down toward his boat and the water line. He could not see any of the bodies that had been there when he left. He turned and walked toward the warehouse were his men would assemble and wait for the promised ride back to Dhaka. He hoped he would never have to get into a boat again. He would be glad to get back to headquarters.

———

March - Denver, Colorado

One of the drawbacks to the notoriety of national politics was that well-meaning assistants and underlings, not to mention political wanna-bes and hangers-on, were always trying to do so much for the politicians that it, at times, became smothering. Andy had first noticed this when he had been elected to Congress in the 1970's but it had become particularly acute during the presidential campaign. He may have been a very busy man, but there were some things he just wanted to do himself.

He had thought that this problem must gall the President and the Vice-President even more, but unfortunately or fortunately he had not had the privilege of finding out first hand. One of Andy's presidential campaign opponents had cut a deal with the power brokers at the convention to fill the ticket, leaving him the odd man out. He then had the opportunity of sitting back for the remainder of the presidential campaign and watch his party's presidential candidate go down in one of the oddest campaigns in history. He had been lucky to be on the sidelines. He had also sat out the campaign for the former President's re-election the next go-round, Andy's Senate seat had been up for re-election and he had won re-election by a moderate landslide. His old friend had taken the vice-presidential nod from the party and together with the new President they had turned the incumbent out of office. Andy often had "what might have been" thoughts about what he might have been able to do, had he run last year, but he was secure in his Senate seat and he was still a young man.

As Andy Knowles walked from the baggage claim in Denver's airport the "smother the politician" routine started again. He had planned on picking up a rental car and driving out to the foothills and NCAR himself, perhaps breaking free early for a trip up into God's country, the majestic Rockies. In fact, he had specifically told his staff not to arrange a ride for him. But, the people in Boulder were trying to shmooze him well and had not checked with his staff, so a brawny young man nearly as tall as Andy himself appeared out of the crowd and offered to take Andy's bags to the waiting U.S. Government sedan. So much for breaking free early.

After initial pleasantries were exchanged with the young man, the first part of the ride up the Boulder Turnpike was carried out in deathly silence. Either the young man had been told not to say anything or he was cowed by the importance of his passenger. Andy decided to break the ice.

"How long have you worked for NCAR?" Andy asked.

"Not long. I was out at University of California as an adjunct professor, and heard GAIA needed someone with an advanced sociology degree and a bio-science background. I had switched majors going into graduate school, so I had both. Been here about six weeks."

"What do they need a sociologist for at GAIA?"

"Well, ..." the young man paused, "I think the answer to that is why Dr. Brady called you. I'm not sure I ought to go into that. I really don't have the big picture."

Andy sensed the young man's unease and said, "No problem, wouldn't want to spoil the big surprise."

Shortly, they left the turnpike, cut across the southern edge of the city of Boulder and ascended the road up to the mesa where NCAR was built. He had been here twice before, once for the first field hearings on GAIA and then for the ground breaking ceremony. He again experienced the awe he had before when he viewed the majestic expanse Denver and the Great Plains stretching out to the east from this mountain top vantage.

The driver had called ahead on a cell phone and so Ken Brady was at the curb waiting for them. Andy was somewhat surprised that only Brady was there. He knew that several super-grade civilian executives of the National Oceanographic and Atmospheric Administration, who were Brady's superiors, were officed on the mesa and he wondered why they were not there to greet the visiting senator.

Andy smiled at himself. He was getting used to the smothering and attention and recognized when it was absent. However, Brady had mentioned the bureaucratic block he had experienced at NOAA. Perhaps Brady was going this on his own entirely.

"Good Morning, Senator. I'm glad you could make it." Ken Brady greeted Andy. "Josh, take care of the car and then come right back up. Conchita needs some help, we want everyone ready for the 10 o'clock briefing"

Josh nodded and drove off. Brady took Andy up the path towards the GAIA building.

"I trust everything went OK on your trip?"

"Just fine, except for the goosebumps. I hope you realize what happens to a politician when someone he respects presents something in this manner, that the bearer of this information could either be the savior of the world or the harbinger of its end."

Ken Brady recognized the Senator's words as both a complement and a warning not to be wasting his time.

"I assure you I was not being rhetorical. On either count." They entered the elevator and went up to the floor where GAIA was quartered.

"What we have planned is a short overview in my office. I will explain the basics and fill you in on the operation we have. At ten o'clock," he sneaked a look at his watch, "my staff will have a full briefing. We also have tele-conferencing ready with San Diego and the Coast Guard folks back in your home state."

Andy had forgotten that he had helped sweeten the GAIA legislation with a Coast Guard facility for GAIA back in New London. Also, he had forgotten what exactly they did there.

They entered Brady's office and Andy was motioned toward the conference table where he put his briefcase to the side of the stacks of charts and documents on the table.

"Coffee?" Brady asked.

"That'd be great. Sugar, no cream. And before we start, I think I'll visit the restroom I saw in the hall."

———

When the Senator returned he saw that Ken Brady had spread the charts out on the near side of the table and placed two chairs for them to sit in. Taking one of the chairs, Andy leaned back and waited for Brady to speak.

Brady also sat back in the other chair and took a deep breath, but before he spoke he pushed himself forward, sitting on the edge of the chair. He locked eyes with the Senator.

"I hope you realize how much we all appreciate your coming here to hear us out. We really do think this is of the utmost importance. We, meaning myself and the other scientists and researchers who run the various parts of the GAIA apparatus, think that the best way to approach this is to give it to people like you who can analyze the best way, politically and realistically, to take action. Scientists are notoriously bad at solving real world problems, we specialize in finding or inventing the problems or solutions and leave it to others to figure out how to implement things.

"The fact is, Senator, that the very reason we all fought so hard to get the GAIA project up and running has come to fruition, in a most frightening and immediate manner." Ken Brady was still staring into Andy's eyes, but neither broke the connection.

Brady continued, "It has taken us the last two years to get all of the elements of GAIA research put together and in synch. Your legislation provided the mechanism through which we can link the research of dozens of different scientists in totally different disciplines together. We put the inputs from weather, ocean currents and temperatures, satellite geophysical data, iceberg sightings, arctic and antarctic snowpack, agricultural forecasts and results, deforestation, pollution, industrial production, acid rain, cloud cover,... a hundred things... together and analyze it in that wonderful big supercomputer your constituents made. The supercomputer itself has a distributive processing program technique which in turn uses the other computer centers who input the data to in turn extrapolate and interpret their own data in the new light of the information GAIA is able to generate from the collected data. Then GAIA again takes this new data from the distributed processors and comes up with a realistic model of how all of the forces of nature, man and the planet are interacting and working on Earth. If you subscribe to Lovelock's theory that the whole planet is a big self-regulating ecosystem, that is, a living thing made up from the constituent parts of all living things, then our GAIA program is the biographer of this living thing, which we portray as the Greek goddess of the earth, Gaia. And our GAIA computer is also Gaia's or Mother Earth's fortuneteller. Because we can see, in a way impossible without the massive body of knowledge the GAIA system gives us, just what happens to the Earth eco-system under given conditions."

Andy was getting impatient to get to the point, "Right, that is exactly what we wanted to do." Andy left the obvious question dangling.

"Well, Senator, we have looked into the billion dollar crystal ball you bought for us, and we got scared to death!"

Brady paused for a moment, but Andy motioned with a rolling motion of his hand to keep the information coming.

Brady took a sip of coffee and continued, "About three months ago we ran the first total global interactive analysis. It was somewhat frightening, but as it was the first run we assumed that the data was skewed or incomplete. We ran it again with double checks as to completeness. When we got almost the same results again, we decided that the main program must be wrong, somehow compressing time or its own calculations. So, we called everyone that had anything to do with the prediction and prognostication software together and came up with a checks and balances scheme that everyone agreed was foolproof. Then we ran everything again, checking every possible place for an error to germinate." Brady gulped more coffee.

"All of the first results were confirmed, with a vengeance!"

Now it was the Senator who was leaning forward in his chair.

"And...?"

"The results were that all of our worst nightmares regarding greenhouse effect, global warming, pollution, ozone depletion, you name it, were really true. And many are coming at us like a freight train, quicker than even the gloomiest prediction could have guessed."

"Such as?"

"Before we get to details, let me give you the really bad news."

Senator Knowles lifted his eyebrows in amazement that what he had just heard was not the "really" bad news. Of course, he had to ask, "And just what is the 'really' bad news?"

"Senator, after we got a true picture of what GAIA was telling us. We decided to see what we could do about it. We created scenarios where fluorocarbons and hydrocarbon thermal pollution and other problems were immediately reduced and then asked GAIA what the result was. The 'really' bad news is that we appear to have crossed the line of no return on some problems and we are frighteningly close to the deadline for other things. The nightmare is already upon us."

Eight people were waiting for Senator Knowles and Ken Brady in the GAIA conference room. Four on each side of the long table. A projection screen was pulled down from the far wall and two small TV monitors were on the wall to one side. The eight stood up as Knowles and Brady came in. Brady introduced his senior staff.

"Senator, this is Conchita Alvarez, our graphic analyst, she's responsible for the graphic images you've just seen and will see more of now." Andy shook hands with Conchita as Ken continued.

"Ted Henke, meteorologist, he is our interface with the National Weather Service and the foreign data banks on weather."

Andy reached up the table to shake hands, it was clear that handshakes would have to stop as introductions went farther up the table in the confines of the small conference room.

"You met Josh Dunleavy, our sociologist." Andy nodded to Josh.

"Midge Carter really belongs to the U.C. San Diego staff, but works here as our liaison on oceanographic matters. Tom Frey is our botanist and agricultural expert. Dr. Tony Wainwright is a chemist, specializing in organic gases and the greenhouse effect. Jenny Livengood and Bart Jastre are both responsible for coordinating the main GAIA software program; Jenny is an engineer and NASA's representative and Bart is a geophysicist and the British entry to the GAIA team." Bart and Jenny were within reach and Andy shook their hands.

Ken pulled his glasses down on his nose and peered toward the TV screens. Seeing them both on, he continued the introductions.

"On the left monitor, you can see Dr. John Colton, my colleague at Scripps Institute in San Diego. You met him during the hearings."

"Yes, I remember." Andy said to the monitor.

"And from your home state and on the right monitor, we have Commander Jeff Cohen, U.S. Coast Guard, our expert on maritime and coastal matters. Can both of you gentlemen hear us?"

Both men on the televisions confirmed they could hear.

Brady stood in front of one of the two remaining seats, motioning Andy to the other.

"The senator has heard the basics from me. Now we need to give him the details." Turning to Andy, "Cut in with questions anytime you want, our goal is to get you fully informed." Andy nodded.

"Ted, you start off." As Ted Henke went to the front of the room, Conchita turned the tele-conferencing camera toward him and switched on the large computer screen that filled the end of the room.

"Although the causes of our current problem are many, the heart of the problems is global warming. We have come to realize that only a change of a few degrees world-wide can have profound effects on world climate and our whole ecosystem. The first slide shows a comparison of the Earth's average temperature during the age of the dinosaurs, the last ice age, the last several centuries and finally on the right over the last several decades and years. You can clearly see the trend towards warmer temperatures in the recent past, especially in the last few decades. And these are just averages worldwide. If we look at specific temperatures, especially at key points, such as Antarctica, mid-ocean, etc. where modest change in temperature can create the most change in the weather, you can clearly see, next slide, that the world is heating up. And as I am sure Dr. Brady mentioned, GAIA has forecast a continued upswing to temperature levels not seen since primordial times and not even present when the last Ice-Age ended and the great world-wide glaciers melted."

Henke continued to present a weather and temperature picture based upon actual readings and computer prognostications. Midge Carter then took over to survey the ocean temperatures and started to explain what effect the higher temperatures would have on evaporation and currents. John Colton followed with a synopsis of what the combined air and water temperatures would mean as far as weather, ocean currents and global icepacks were concerned.

One by one the panel went around and filled in the myriad of pieces that made up the puzzle which the Geophysical Atmospheric Interactive Analysis system was meant to decipher. Gaia gave up her secrets to Senator Knowles.

At noon Brady's secretary brought in sandwiches and soft drinks. The discussion in the room continued.

By mid-afternoon everyone needed a break. The Coast Guard officer in the East was released from the tele-conference due to the local time there. He had presented his part on coastal conditions and effects on maritime traffic of the foreseen events. The Senator let him know he would be stopping in at New London on his next trip home.

After they reassembled from the break, Brady spoke first.

"Senator, we have pretty much covered the situation we believe is facing us. Now we should turn to what we see as a plan of action, insofar as one presents itself. And as part of this, we need to again touch upon why we called you into this. Perhaps that should come first.

"Our GAIA project is, of course, administered as a part of the Department of Commerce, through N.O.A.A., but our staff comes from many disciplines and organizations; federal government, state university, military, British Academy of Sciences, you name it.

"When we started coming to the realization of what a bombshell we were sitting on, we also started putting out feelers as to how the information would be received and how it should be dealt with. We are part of the Office of Climate and Research at N.O.A.A. and they work very closely with us and coordinate things with other branches of the government, but this is the kind of thing that needs coordination at the highest levels of government. When we first took the results to N.O.A.A. they actually told me to be careful about acting like Chicken Little. They told us to submit a complete report for review.

"Three weeks after we had sent it in, I checked on its status. They hadn't even touched it, except to schedule it for presentation to a symposium of meteorologists next September. When I tried to raise a ruckus at N.O.A.A. I was quickly informed that it was departmental policy to fully assess new scientific information through 'proper' channels. They indicated that both the White House and the Commerce Department were analyzing our report and would give the "hypothesis" a thorough review. A senior staffer from the Under-secretary's office even called me to let me know that it was Administration policy to handle environmental policy at the highest levels and that they did not want any public release of doomsaying about the environment that was not cleared by the top at N.O.A.A., especially in light of the President's top priority of keeping the economy on track."

At this point Andy got up to walk to the large world map at the back of the conference room, he indicated with a roll of his arm that Ken should continue.

"We assumed that because the Under-secretary's office and presumably the White House and E.P.A. was involved that this was being looked into at the top. However, when the Under-secretary came through Boulder last month, I had the opportunity to talk to him at a reception. When I asked him what he thought of the GAIA information, he said that it was an intriguing theory, this global warming idea, it would be interesting to see if anything panned out over the next decade to support it. I tried to tell him something of the urgency we felt, but I was told, in so many words, that it was the President's policy to thoroughly assess all of our alternatives and continue on our policy of environmental concern and addressing known problems in proven ways, while looking for new ways to deal with problems also, but that he had to be careful and not throw the baby out with the bath water by doing anything that could set back the economic recovery. He clearly thought I was a nut case. Worse yet, he really did not have the gist of what we were saying. He acted like the warnings of global warming, ozone holes and what not of the last two decades had been the boy crying wolf and he didn't want to hear anyone crying wolf any more, at least not while he was in office. We made other inquiries through our various channels in other agencies, but we almost universally got the word that from the cabinet level, the word was to tone down environmental problems, they feel the economy needs good news to get on its feet and that the environmental problems can be dealt with through existing programs. I was shocked that with the President's, and certainly the Vice-President's, well-known positions on the environment that this would be Washington's position, but I was told that the President had been elected because of the bad economic climate under his predecessor and he didn't want anything to set his vaunted economic program off course. In short, we were told that the people at the high levels who we needed to see the information GAIA had given us, did not want to see any bad news, at least not now. The political appointees at the highest level of the Commerce Department wanted the focus on restoring the economy, and wanted nothing to interfere with the process.

"We thought about our other options. Having Bart do a paper for the British Academy. Or just calling a press conference and laying it all out for the press and the people to decide. But, this is so catastrophic, earthshaking, God knows what the right word is, that we all decided that it had to be handled in just the right way. If we did have a press conference and they believed us, then there could be world-wide panic, economic collapse, who knows. If we just issued another of the dozens of scientific reports talking about the greenhouse effect, it would probably have so little effect that it would be a useless exercise and it certainly wouldn't initiate the level of action we think this needs.

The need is obvious. The problem is obvious, at least to us, and hopefully after today, to you. The problem is that we need, the world needs, a plan of action from the very start, initiated at the highest levels.

"So, that leads us to you. We all took stock of our political clout, and quickly realized that we are scientists, not politicians, and the only a few people at national levels had ever seemed to take the GAIA concept seriously,including you and the Vice-President. Besides, your campaign rhetoric from your stab at the White House indicated to us that you really believed that a national environmental policy, along with a national energy policy was as necessary as a defense policy. I think the Vice-President and hopefully the President, too, feels that way, but they are being insulated by well-meaning, but mistaken staff members."

"So, you think I can do something with this or at least get it through to the top, if you put it in my lap," Andy asked.

"We're praying you can."

Andy Knowles thought for a long time before continuing. No one spoke, as it was obvious he was getting his thoughts in order.

"OK, let's hear what you think ought to be done." Andy started to sit down again, but paused and turned to Bart and Jenny. "But first, let me ask a few questions." He cleared his throat and said, "What are the chances that the whole GAIA prognosis is wrong, just a computerized over-reaction to a few hot years and unique phenomenon?"

Jenny spoke first, "After our own first reaction to the magnitude of the problem we instituted a self-evaluation program in the GAIA computer matrix. It gives us a statistical probability of each level of prediction. That is, in lay terms, it predicts the probability of any given thing happening right along with the best and worst case scenarios. Regrettably, the possibility that this is a mere anomaly that GAIA is overreacting to is so small that it is statistically impossible. Besides, the evidence is so overwhelming and broad-based that even from our own personal perceptions we can see clearly that GAIA is on the right track"

Bart Jastre cut in, "Senator, we have asked ourselves the same question that you ask us now. The very frightening answer is that it is statistically more possible that the greenhouse effect will go totally out of control and destroy life as we know it, than for things to continue as usual."

Seeing Andy Knowles' worried reaction to his last words, Jastre put up his hands and quickly added, "Please realize that that was just for argument's sake. Neither normalcy nor total end of the world possibilities are within the range of real consideration."

Jenny jumped back in and continued, "Commander Cohen used reasonably possible best and worst case scenarios when he talked to you about the effect on the coastlines. And there is a point about a third of the way between best and worst cases that is statistically most probable. There are charts in the materials we prepared for you to take that show best, worst and most likely scenarios for all the major factors, weather, sea level, populations effected, etc. The only real question we have left is one of time. This is such a delicate balancing act that a few factors, one way or the other, could make certain events occur in years, decades or centuries. Also, there is the principle that former Vice-President Al Gore wrote about in his book where one factor out of many possible finally tips the scales and the whole system is reset to flop into a new level of equilibrium. And, of course, what we do in reaction to these events also effects timing and can effect the future's best, worst and most likely scenarios."

Ken Brady now spoke, "What GAIA has taught us is that life on Earth as an ecosystem, a living thing in and of itself, is really a balance, GAIA can tell us how far the balance beam will swing when certain weights are added to or removed from the scale and, more importantly, tell us where it will re-balance itself to a new equilibrium once it does swing. Our industrial society has been adding weight to one side of the scales for two centuries now, in the form of pollution, deforestation, energy use, chemicals, and simply the wasteful way we live. Now, GAIA is telling us that the balance beam is starting to swing. There is no question of that. The only question is how far and how fast, and where it will reach a new stability. And that depends on what we can do to help Gaia, the concept of life as a self-balanced system, swing back toward the norm or reach a new sustainable level we can live with. GAIA, the computer system, has merely warned us that Gaia, the Earth Goddess, Mother Earth, is really upset at what we have done to her creation with our dirty, wasteful lifestyle."

Senator Andy Knowles leaned back in his chair and looked around the room, "Then I guess you had better let me know what you think we can do to soothe Gaia's hurt feelings."

———

Senator Andy Knowles was the only first class passenger on the red-eye flight back to Washington. With the time difference it would get him in to Dulles Airport well after midnight. The long ride in from Dulles would mean he would not get home to Bethesda until two, maybe two thirty.

Luckily he had been able to get a call through to Derek and to his wife before he left Denver. Molly would meet him at Dulles. He had protested that it was not necessary, but she insisted with her usual retort that she could not get to sleep while waiting for him to get home.

Derek had been given instructions to clear as much of Andy's calendar for the rest of the week as possible. And he had been told to have the staff pull together "the works" on the Commerce Secretary, who was known as being a politician first and an administrator second, his under-secretary in charge of N.O.A.A., and everything known about global warming, with an emphasis on the administration policy on global warming, greenhouse effect, fluorocarbons, etc.

Andy spent much of the flight sifting through the stacks of charts and reports the people in Boulder had given him. Although the dire predictions the scientists had made kept coming to the fore, Andy made sure he concentrated on getting a clear picture of the current evidence they had put together. The hard facts were what he had to master because that was what everything else was based on. And it would be the hard facts that he would need to get things moving in Washington. Once he had convinced them of the facts, then and only then could he scare the shit out of them with the visions from the crystal ball that the people of the GAIA project had provided him.

But knowing that only left the bigger questions, and those questions were the ones the scientists counted on him to solve.

How should he handle the predictions of GAIA?

How should he tell them? Obviously, if he could get the White House involved that would be best. He had a hard time believing that the two leaders at the top, especially the Vice-President, had turned a deaf ear to the environmental nightmare GAIA had warned of. He had worked with the Vice-President on many issues of ecology and environment in the Senate and he knew how fervently the woman believed in the environmental issues. He suspected that Brady was right in assuming that the intermediaries were blocking the information from getting to him. He could imagine how the top Commerce Department officials might fear giving information like he had in his briefcase to a known environmental zealot like the Vice-President or to the President who had vowed to make the environment a cornerstone of his administration's policy. The nightmare that Gaia warned of could throw the economic programs so dear to the Commerce Department into a tailspin.

And, most importantly, what could be done to insure that GAIA's horrific scenarios for the future were blunted, nullified or, God forbid, merely postponed for as long as possible?

As he studied the facts and prognostications again and again, Andy came to fully understand the problem the scientists faced. A simple press release could have as little effect as a dozen other forecasts of global warming had had over the last twenty-five years. Or if it was done with too big a splash and without the proper follow-up and control it could have immediate repercussions in economics and society which would do little or nothing to solve the problem and would do great harm to everyone concerned. And, of course, the worst outcome of this would be an overreaction to the initial news followed by retrenchment and inaction. That was especially dangerous since they did not have a good handle on the time available to them to take action. The "boy who cried wolf" situation was entirely possible here. If it did not already exist from the callousness much of the world showed to the previous greenhouse effect and global warming news. No, what was needed was a cogent, organized policy from the highest levels. And it would, of course, need support from not only a broad spectrum of Americans, but from the whole world.

Andy Knowles found himself in awe of the situation he had spread before him. For a brief moment he reflected on the fact that perhaps he was in one of those fulcrum points of history, in which one man's efforts made a difference in the world.

As his plane circled down towards Dulles International, Senator Andrew Stuart Knowles struggled with visions of what might be and what could be. And he put together a plan of action to have the world take note of Gaia's complaints.

———

March - St. Petersburg, Russia

The old man walked leisurely through the bustling crowds of the city street. The office workers were taking advantage of the beautiful weather of this early spring day to run noon-time errands and take brisk walks in the welcome sun after the long Russian winter. The last sooty piles of plowed snow melted in the sun.

Many of the passers-by nodded greeting to the old man and instinctively made way for him to pass. Those who were not familiar with the silver-haired and square-jawed good looks of the man also showed their respect when they caught a glimpse of his lapel. His looks and his gait belied his ninety-plus years.

A Hero of the Soviet Union was not a common sight these days. Most of those who held the old Soviet honor kept the medals from view in the modern Russian Republic. And this old man had one of the revered red stars suspended from a tattered ribbon on his chest. Most citizens of St. Petersburg, and in fact most of his countrymen, knew of the exploits of the man who, as a young boy, had been honored for his heroics in defense of the city of Leningrad, as St. Petersburg was then called, during the Siege in the Great War. Nicolas Litasku`s biography had been standard reading in the Young Pioneer's manuals throughout the former Soviet Union. To have an ethnic minority, an Estonian, as such an important hero of the nation was considered an important part of the education of Soviet children, both minorities and Russian. Now that Estonia was an independent country and Soviet propaganda had disappeared behind Russian nationalism, his story was not retold as fervently as it once was, but still many people knew of him, especially in St. Petersburg.

So, some of the people who passed the man felt they knew him. In year's past and occasionally even now, children would beg him to tell the story of his exploits. But today was a school day and there was no one to stop the very old man. An occasional police officer would touch the brim of his cap as the frail old man passed. For "Nikolasha" was not only a Hero, but his son was also the commander of the Militsia for the city and the police officers knew the old man had the ear of their boss.

From the crest of the old stone bridge, Nikolasha viewed his favorite part of the city, the waterfront. To the left was the wharf where he, barely a teenager, had taken a job as a stevedore. It was from there that he and a hundred other men and boys had been formed into a naval infantry company to defend the marshlands and seacoast west of Leningrad from the Nazis. Only five of them survived the battles that miserable winter, the battles with the Nazis and the battles with the bitter cold. He, himself, more often thought of the events in terms of the embellished stories he had heard so many times since, rather than the faded memories of the real events. He had been all of twelve years old at the time of the exploits, the actual memories were dim.

More clear in his mind were the memories of the years he had spent in this port, his wartime exploits rewarding him with the plum job of Harbormaster at a very early age. Before his retirement, now nearly three decades ago, he had been the voice of authority for everything that went on around the Leningrad waterfront. Even now, from his honorary seat on the Board of Advisors, he could make his presence felt.

The Neva River, which flowed through the city, creating the harbor, was running high this day. It was odd that it was so high, usually the Neva only rose this high when a storm to the west in the Gulf of Finland or the Baltic caused a tidal surge and it was too early for the melting of Lake Ladoga to be causing a flood. Surely, with such fine weather here there could not be too large a storm to the west. It was probably just a very high springtide. He gave thanks to the god of his childhood that such a lunar tide had not combined with a storm. Perhaps thirty years before, this had occurred and the Neva had shown itself able to flood much of Leningrad. The combined flooding and storm had made for many long hours for the harbormaster. The only land that Peter the

Great had been able to steal from Sweden in the 1700's had been tidelands and marshes between Lake Ladoga and the Gulf of Finland and upon that he had built his great city, St. Petersburg. The old Communists had renamed the city after Lenin, but they could not do anything about its geography. And even the name of Leningrad had not endured, having been voted down by the people in the wake of Perestroika and Glasnost and the end of the Soviet Union.

With nothing better to do until he picked up his granddaughter at school that afternoon, Nikolasha continued to stroll down to the new automated docks. The overhead gantries and container movers were greatly different from what he had used in his day. But the whole operation was so different. When he had first returned to the harbor after the Great War, many of the ships in the port were old American "liberty" ships, leftovers from Lend-Lease programs during the war. Throughout most of his tenure in the port the ships remained predominantly Soviet or Warsaw Pact countries. Now you could just as easily see a ship from Japan or Liberia as a Russian ship. The longshoremen, stevedores and ships crews were busily at work loading machinery and farm products, taking advantage of the weather, they knew that a spring-like day in March was no guarantee that winter would not appear again tomorrow.

As Nikolasha walked along the upper quay wall he heard a commotion from one of the cargo trans-loading platforms below. He made his way down the iron stairs with a spryness that belied his eighty plus years. Reaching the bottom he heard a burly man with a clipboard bellowing at an assembled crew. "I leave you alone for a few hours and you ruin fifty pallets of grain. And then you don't see fit to tell me about your stupidity until I find it myself when you leave for break." The burly man, known to Nikolasha to be a foreman for the granary loaders, was red-faced with anger.

A blonde man with a hint of Estonian accent spoke up. "We had no idea it would be so high a tide. We put the pallets where we thought you expected them. It is not our fault that they were soaked during the night. If you thought it would be a high tide, you should have told us."

The foreman started to spew forth a string of obscenities when Nikolasha stepped forth.

"He is right. In my day, it was a foreman's duty to check the tide tables at the harbormaster's office and stack the cargo above tide level."

The man whirled in anger, but promptly checked himself when he saw whom he faced. "I did check the tables last week. We don't have a dangerous tide for the rest of the month and it is too early for runoff from the Neva."

"Then why do you yell at your men. A foreman's job is to report the damage to the office and not act like a buffoon and try to place blame where none should be." Nikolasha locked eyes with the big Russian for just an instant before turning and climbing back up the stairs. As he went up, he noticed the water line on the pier where the water had risen to the night before and this morning tides was almost up there again. It was a good four feet above the normal high tide. He would have to check the tide tables himself at the library. It was most unusual to have such a tide in calm weather, the lunar charts must have something unusual. That dumb Russian had probably made a mistake.

As Nikolasha continued his stroll, he surveyed the bustle of the waterfront. The newer piers that had been added to the west had greatly increased the port capacity of St. Petersburg, but everything still had to come through the warehouses and yards surrounding the railroad yards behind him. The increases in foreign trade that had come with the breakup of the old Union still had to be funneled through the handful of ports.

On a nearby pier, in contrast to the clamor of cranes and trucks on neighboring piers, Nikolasha noticed an odd lack of activity. The talent of the old harbormaster to spot a problem in an instant was still there. The lack of activity in such a high priority location has troubling. After a moment viewing the situation, Nikolasha saw the problem. The ship, a German cargo container carrier, was tilted crazily in toward the pier and from such an angle the cargo container handling machines could not pick up the containers properly. Nikolasha could not resist going down and to see what was going on.

The security guard at the head of the pier saluted Nikolasha as he walked down the pier. Nikolasha could see by the red truck parked near the ship that an assistant harbormaster was already on the scene. Whereas Nikolasha had done the job of Harbormaster with only the aid of a few men, his successor now had a dozens of young fools trying to keep things running smoothly. From his seat on the Board of Advisors Nikolasha had made known his thoughts of having the young "maritime engineers" put in charge of port operations when they had no experience on the waterfront. Nobody had listened.

As he approached the problem became abundantly clear. The graying nylon lines anchoring the ship to the pier were taut and ready to break. Nylon was funny that way. It would stretch far more than the old manila lines, but when it reached its breaking point it was made clear by the discoloration, noise and heat from the friction inside the artificial fiber lines. The worst danger was that when a nylon line parted itself, it would whip away at supersonic speed, like a bullwhip. Many a seaman and stevedore had lost limbs and lives to snapping nylon lines. Worse yet was the fact that the typical hitches taken in shipboard lines around cleats and bullnoses on the piers made it nearly impossible to relieve pressure in the line. The nylon under stress deformed into itself and made a knot that was impossible to untie.

Nikolasha walked over to where the young assistant harbormaster was staring up toward the ship. Clusters of dock hands were standing around in nervous conversation. As he walked up, one of the rubber bladders used as fenders between the ship and pier was pulled by the force of the rising ship up above the edge of the pier and the ship hit the pier with a gnashing scrap. As the ship lurched over, the forward nylon lines stretched even tighter and began smoking, the last warning from the nylon before it would pop.

"You've got quite a problem here." Nikolasha said to the back of the younger man. "Didn't anyone tell you about high tide when you became a 'maritime engineer'?" The sarcasm in Nikolasha's voice was quite apparent. The younger man did not have to turn to recognize the gruff old Estonian by his accented Russian.

"Gospadin Litasku, we still know all about high tide and the precautions were made. Proper slack was put into the lines at low tide and the fenders were right for the tide charts and ship's draft marks. But, if you will note on the water line marks, the water is four feet above mean high tide. Much higher than expected. The men were in a hurry to unload this food cargo and the extra high water plus the lessening of draft of the ship when it was unloaded quickly at high tide combined to draw the lines tight and now we can't get them off the bullnose."

He was at the side of the young harbormaster. "Have you tried to reload the ship and push it down?"

"We thought of that, but the containers were loaded on trucks and away before we could get them back."

"What are you going to do then?"

"I don't know!" The young man had started to yell at the old man, but had tempered his tone and continued. "If you have any ideas, we would appreciate them."

Nikolasha had nothing to say. As he watched, one of the smaller lines amidships parted with a rifle-like snap. All of the onlookers on the pier and shipboard moved back farther from the remaining lines. A trash can that had been in the path of the line's arc of recoil had been smashed and knocked into the water on the other side of the pier.

Nikolasha Litasku cupped his hands and yelled in German to the ship's captain standing on the bridge wing. The captain considered what Nikolasha had said for a moment and turned to go into the pilot house with a jaunty salute of acknowledgement.

"What did you say?"

"I thought you 'maritime engineers' were supposed to speak foreign languages?"

"I speak English and French." the young man defended.

"I just told him to take on ballast and flood his bilges. You do know what a bilge is, don't you."

The younger man nodded and struggled to keep the look of relief off his face. The old man's suggestion was obviously right and would probably save him from having to make excuses for injuries or damaged equipment.

As they watched they saw the ship settle down in the water ever so slowly. Nikolasha turned to go.

"Thank you, Comrade Harbormaster." the young man said as he turned. This time the young man intentionally used the old `Comrade' instead of the `Gospadin' or `Mister' that was now in vogue in post-Soviet Russia.

"You are welcome." Nikolasha grinned slightly as he turned his back on the young man. Experience still had some value.

As he walked up the pier, Nikolasha paused to once again look over the side to the water. He shook his head in amazement at the high water line. He would have to check the almanac on this. The weather was clear, no storm, no low pressure system. Why was the water so high?

———

Nikolasha was sitting at the dining table when Sofia came home. The television was on; she was late, "Vremya" was already on. She could hear her daughter, Daine in the other room. The old man and the young girl had a running competition of turning on "Vremya" and rock music in different rooms each evening at volumes loud enough to drown each other out. But, this evening both seemed subdued.

"Istvan is not home yet?" Sofia asked. She made a point of speaking in Estonian, the old man usually warmed to her attempts to speak his native tongue. Deep down she knew that the old man knew how hard she tried to please him with such efforts. It had started shortly after her marriage to Istvan, when the old man had made it known he was not happy about his only son marrying an ethnic Russian, especially a physician from Moscow. Knowing that, under the restrictions and cost of housing, she was destined to have the old man in her home until he died, Dr. Sofia Litvinov Litasku had made the effort to learn her father-in-law's native language, and she even condescended to naming her daughter after the mother-in-law she had never met, Nikolasha's Estonian wife and Istvan's mother, Daine.

This time her attempts to charm the old man were fruitless, in his most guttural Russian he responded, "No, Istvan is not home, he is probably still downtown kissing Russian bureaucrat's ass."

Sofia never let the old man get away with snipes at her husband, "It is very fortunate that in your career as Harbormaster you never had any occasion to kiss Russian ass. Is it not?" She knew that the old Estonian who had risen as high as he had in the Russian bureaucracy would have no response to that.

Changing the subject, she said, "Did you hear about the flooding? We had three patients, injured in the floodwaters?"

"Yes, I saw it, yesterday and today. I was hoping to hear something on the news program, they had some flooding on the Volga and Don, but did not report anything about the Neva or Lake Ladoga."

Sofia had been shopping, a daily occurrence for many Russians as they heard what items were for sale on a particular day. Things had gotten better in the last few years, but they still kept the habit of hoarding and bargain hunting. Years of living at the edge of the abyss in the trying times after the falloff the Soviet Union were hard to forget. She set to work putting things away in the kitchen. Nikolasha continued to watch the evening news from his seat at the table, surrounded by stacks of books and the black binders Sofia recognized as his journals.

Istvan finally opened the door to the flat shortly after eight. Sofia had already served Nikolasha and Daine their evening meal. Daine had taken over the television and was watching her favorite variety show. Nikolasha was still thumbing through his journals at the table.

Istvan's normally impeccable Militsia uniform was rumpled and dirty. He had muddy water lines above his knees and his jack boots were covered with dirt and mud.

Sofia met him just inside the door with her usual hugs and kisses. Istvan returned the affection which was cut short by Daine intervening for her own hug.

After the embraces of his wife and daughter the hug and cheek touching with his father was subdued, but no less meaningful. The father was proud of his son and the son dearly loved the old man.

Nikolasha pushed Istvan back to arms length and surveyed him. "Commander Litasku, you are fucking mess." Nikolasha laughed. "You have been consorting with Russians again."

Istvan laughed too, but took a nervous side glance at Sofia. She was in the kitchen. Istvan was preparing to sit in the overstuffed chair near the dining area when Sofia came rushing in, clucking," Ut-tut-tut-tut-tut!" She was carrying the two large plastic bags which her purchases for the evening had been bundled in. She put one in the seat of the chair and the other at its foot. She then motioned Istvan that he could now sit. As he sat, Daine rushed over to him to carry out the nightly task of helping Papa off with the tall brown boots. She hesitated when she saw how muddy they were, but went ahead anyway. At her mother's direction she took the boots and the socks under them into the kitchen to get cleaned up.

"I see the flood found you." Nikolasha said as Istvan finally settled back with a sigh.

"In places everything on both sides of the Neva is under water for several blocks. I hear that the whole area to the east towards Ladoga is under standing water, luckily it is out of my district."

"Any serious damage?"

"Nothing too bad, the usual wet basements and power outages. You'd think the utilities would be better prepared. We would have had a mess at the central archives building. but they already had the lower level cleared out from the broken water pipes last month. I got this," Istvan indicated his muddy legs, "Because some stupid Central Asian bus driver thought he could cross Nevskij Prospect with a school bus when it was under water. By the time he got stopped he was in a meter of water. Flooded the engine's electrical. The school children had to perch on top of the seats until we could get them. He said he couldn't read the deep water warning signs because they were partially submerged, all he could see was 'caution,' so he drove in cautiously. Bozhe Moi!"

Istvan and Nikolasha laughed heartily. The immigrants from Central Asia to the big cities in Russia were always a good brunt of jokes. Sofia interrupted by bringing Istvan's warmed-up dinner in. She unceremoniously brushed Nikolasha's books aside.

Stacking his journals together, Nikolasha spoke, "I was at the docks again this morning. It was chaos. The water is higher than anything I can remember."

"I know. Before I left the office this morning we got a telegram from the Interior Ministry that predicted this to continue for two weeks. It seems the snow in the countryside is melting earlier than normal and a low pressure system near Sweden and Denmark is combining with the tidal forces to make the Gulf of Finland at high tide stage for an extended period. The result is that the Neva and its tributaries are dumping a lot of snow melt and the outlet to the sea is higher than normal, so we get flooded."

"Ahh, that would explain it." Nikolasha nodded his head in understanding, apparently the mystery was solved for him.

Istvan wished he could brush the flooding off as easy as his father had. Of course, he had not told the old man everything. There was no need.

Commander of Militsia Istvan Nikolaevich Litasku had received another telegram that morning. Moscow, the Interior Ministry, had ordered all local officials to relocate all critical communications and emergency supplies to areas of at least three, preferably five, meters above sea level. They were worried about a continued increase in sea level that had been detected over the past six months. It was not just St. Petersburg that was having problems. The great rivers of Siberia were out of their banks from runoff and the Black Sea resorts and cities were having difficulties with high tides.

Commander Litasku had replied to Moscow by telegram that the Central Leningrad Militsia Oblast had no land areas which had such elevations. Only the highlands to the north and southwest were that high, and these areas were fifty kilometers outside of the city. All of the city proper, his area of responsibility, was lower. Moscow had told him to do his best, to insure that his assigned watercraft were well-maintained and provisioned for rescue and patrol duties and await further instructions on relocating critical materiel.

All local police and security officials who had received the warning and emergency orders were also directed to keep the entire matter secret including from the local civilian government which nominally, in normal times, was whom the militsia answered to. The classification caveat indicated that the orders were from the highest levels of government.

———

March - Washington, D.C.

The late night had not stopped Senator Knowles from surprising his staff at the stroke of eight. Only the receptionist and Derek were in the office when Andy came in. The schedule of Congress in session made the members somewhat nocturnal in their habits. The mornings were reserved for necessary media events and hearings, with the real serious legislative sessions and bulk of the work starting after mid-day. To have a Senator, especially after a late night flight, pop in on them at eight sharp was a surprise.

"Good Morning, Senator," Christie, the receptionist, said as she slid a nail file into her desk drawer and got up and went over toward the coffee pot in the anteroom. The Senator always expected the coffee to be ready and she had not even thought of it yet this morning.

"Good morning," Andy nodded as he passed her and went directly to his office.

Derek started to speak also, but was cut off by Andy.

"Derek, how we doing on the things I gave you last night?"

Derek was looking at his watch and talking as he came in, "Christie is already working on clearing the schedule, but almost no one is in yet. Both the Secretary and Under-secretary were confirmed by the Commerce, Science and Technology Committee. I have Jody waiting at the committee staff's door waiting to get a DVD of the file when they open. I'll put Bill and Dee Dee, she's the new intern, started yesterday, on the global warming thing when they get here." Derek stopped to catch his breath and looked his boss and friend in the eye. "Would you mind filling me in on what's going on?"

"You know the gist of this global warming thing I'm looking into and suffice it to say that I hit pay dirt and its too long of a story for us to go into right now. You'll catch on as we go. First, I've got some other things to do. I want you to make sure everyone keeps this whole thing close to the chest. Second, when they get in I want you to check with the Whip to see if he can figure out how I can get to see the President. They'll want to know why. Just give them some line about how you are not sure, it's some kind of personal thing the Senator Knowles wants to discuss with the President, but doesn't want to go through the White House hierarchy."

Derek was staring at Andy with raised brow now.

Andy continued, "If that doesn't work, nose around and see if anybody close to our camp has an inside track to the President's appointment people. I'd prefer not to go through the White House Chief of Staff, if I can help it. But that may not be possible."

"Why are we trying to get into the White House and avoid the Chief of Staff?"

"Aren't he and the Commerce Secretary fellow travellers, so to speak, on the economic program? I heard they have been getting together to play hard ball and weed out opposition to anything that might derail the post-Covid revitalization program they came up with?"

"Yeh, I think so?"

"Well, I think the Secretary and/or his one of his underlings are shaping up as the villain in this story, and if he's not I think I'll try to portray him as one, and I don't trust the Chief of Staff not to be privy to what's up with him or not to warn him of what I plan to do, or say to the President."

Christie came in a stack of messages and mail. She took Andy's coat from where he had flung it on a chair to the coat tree in the corner.

Andy spoke again, "Christie, call Sylvia Janaczek and see if her brain trust can come up with a copy of the President's campaign platform on the environment or any speeches he made during his campaign or the old gubernatorial campaigns on the environment, and maybe the keynote address from the convention the election before last. Don't let on that I asked for it myself, just say one of the staff is doing research. Derek, you get going on your calls. I have several to make myself and then we can get our heads together and fill you in on this over some coffee and, Christie, get someone to get some of those cheese Danish we had last week. OK?"

"Anything you say."

Derek was mildly annoyed at Andy's secrecy, but realized from long experience the way Andy worked when he was in this kind of mood. It was just that he had not seen Andrew Knowles in this kind of mood since the latter stages of the presidential primaries years before. Senator Knowles was obviously a man with a mission.

———

Christie preceded Derek into the Senator's office. She carried a tray with the pastries and two coffee mugs. She had to sit the tray on the credenza, as Andy had the desk covered with charts, maps and bundles of reports, his laptop open in the middle of it all. Derek took a seat in one of the armchairs facing the desk.

Christie paused for a moment looking for any acknowledgement from Andy. Andy had his feet up on a side chair looking at a colored chart, facing away from her.

"Anything else?" she finally asked, jarring Andy from his thoughts.

"Er, ah, no. Thanks." Andy finally said after spinning around and seeing the tray.

Christie left and Derek leaned forward to look at the piles of information spread out before Andy.

"I've got Bill and Dee Dee both over at the Library of Congress,..."

"Dee Dee, who's Dee Dee?"

"Dee Dee Lindquist, I just told you, she is the new intern, she started yesterday when you were in Colorado. She's on the federal grad student internship program, the President Pro Tem assigns them to senate offices and it was our turn, so we have a new intern. I'll give you her file if you want. Anyway, I have her over with Bill at the Library of Congress doing the global warming research, but it looks like you already have a hell of a lot here."

"Yes, indeed, probably more than I ever wanted to know on weather and floods and storms and geophysics and you name it. The scientists in Boulder really spilled their guts, and their hearts, yesterday."

"Well, I assume you'll get around to the story soon. But for my part, I've struck out. Nobody I can trust to keep quiet about it will admit to having any good contacts in the Executive Office of the President. And everybody else says that all schedule matters are coordinated through the Chief of Staff and that includes meetings with members of Congress. Even the lesser cabinet members go through the Chief's office for meetings and only a select few of the cabinet even call directly."

"I got something of the same thing from my calls." Andy said as he reached over and took a pastry.

"Judy took over from Christie on clearing your schedule. She wasn't sure the meetings on Friday could be canceled without a big tiff back home."

"Yeh, yeh. Just have her do the best she can." Andy was impatient at having to deal with the more mundane aspects of politics, a trait Derek recognized from way back.

Andy took a gulp of coffee to swallow his pastry and reached for the phone. "If all else fails, try the quarterback sneak, up the middle."

Derek watched as Andy dialed the number 4-5-6-1-4-1-4. The White House. Andy punched the speaker phone button and re-cradled the phone.

"The White House"

"Office of the President ,please." There was a pause.

"Executive Office of the President, may I help you?"

"This is Senator Knowles, would it be possible to speak with the President?"

"I'm sorry Senator, the President is in a meeting and his schedule is somewhat full. If I might suggest that you contact the Congressional Liaison Office. They handle all matters for the President and members of Congress. I can connect you."

"Thank you, that won't be necessary. I have that number."
Andy thought of just leaving a message, but he realized that it

would not get to the President, rather the Chief of Staff or the Congressional Liaison people would get it and return his call. Andy hung up the phone.

"It's a little ironic. A couple a thousand more votes in the primaries and I might have moved to the White House and now I can't even get through on the phone."

"That's true, but you have to admit, we do the same thing. There's even lots of times when other members of Congress try to call you and I return the call or Judy does. One politician's important call is another politician's annoyance. And that's especially true since you don't want to go through channels. And... he is the President."

"Right, right. Well, there is still the end around play." Derek smiled at Andy's continuing talk of football plays. Andy was fond of his football days at Yale and often related to them in his conversation. They had talked to him about it in his speeches more than once. It had become a point of derision for some journalists during campaigns.

Andy took a black notebook from his top drawer and thumbed to a page. He looked up a number and dialed. He again put it on speaker for Derek to hear.

"Office of the Vice-president," was the answer.

"Sarah Vaughn, please." Andy said. Derek smiled when he recognized the name of the woman who had been the office manager for the Vice-president when she had been a senator in an office next to Andy's. She had gone with her to the Executive Office Building when she had become Vice-president.

"Hello."

"Sarah, this is Andy Knowles. How are you doing?"

"Just fine, Senator. It's been a while."

"It sure has. Sarah, I need to talk to your boss. Is she in?"

"No, she's not. She's spending the morning at the Admiral's Quarters. Getting ready for a trip to Kentucky and Tennessee."

"Can you get me a number where I can reach her? I need to speak to her before she leaves."

"Certainly, Senator."

Sarah Vaughn gave Andy not only the main number to the Admiral's Quarters, the official residence of the Vice-president, but also a private number that would get through directly, from the area code and phone exchange, it appeared to be a cell phone. For Sarah Vaughn remembered Andy and the Vice-President had been close friends and confidants in the Senate in the days before both had become enmeshed in presidential campaigns and Sarah thought nothing of giving Andy her number.

———

March - Dhaka, Bangladesh

The wait for a ride and the ride home took most of the night. None of the buses were running, but Captain Darpheel had managed to secure a spot on an Army truck headed for Dhaka for himself and his men.

The truck was one of the awkward looking Japanese cargo trucks that had come in an aid package. It had dropped a load of supplies off in Narayanganj and was headed back for the airport in Dhaka. Akim Darpheel and his sergeant sat in the cab with the driver, another Army sergeant, and Darpheel's men rode in the back.

Dead tired from their journey, Akim and Quist tried to rest their heads on the window behind them, but the road was too rough. Normally the road from Narayanganj was an excellent ride, by Bangladeshi standards, but the floods and debris had wrecked the road. Now, the ride which would be a simple ride in from the suburbs in another city and another time had become a journey of several hours. Several times they had to detour and once they all had to dismount for the truck to cross a particularly questionable pontoon bridge that had been hastily erected to replace the ruined bridge.

"How badly was Dhaka hit?" Akim had asked the driver.

"They say the typhoon went straight up the Brahmaputra valley and the eye of the storm passed just to the west of Dhaka. The Burhi Ganga is out of its banks. Everything is ruined there. Some of the big apartment buildings downtown, the concrete ones, just blew over. And many of the other buildings are ruined. All of the wooden houses on the south-east side and over near the airport were flattened. I saw them on the way down here," the driver said.

At this, Sergeant Quist took in a breath and muttered a prayer. His family lived near the airport.

"I spent the night with my family in the Al-Akhbar Mosque. The wind shook the building and some tiles fell on our heads, but the mosque came through." The driver spoke with the annoying habit of taking his hands off the wheel to gesture as he spoke. "A lot of others were not so lucky. They are burying people in trenches dug out with tractors all around Dhaka."

Dawn was breaking as they drove into the Army motor pool near the airport. Darpheel had finally managed to catch some sleep and awoke when the truck jerked to a stop. Darpheel and his men would have to finish their journey by their own means.

As they climbed from the truck, Darpheel realized that the men would be wanting to check on their families. They had done enough for now.

"Everyone is dismissed for now. I want everyone to report back to their normal assignments tomorrow morning at headquarters. If anyone is unable to get downtown or has other problems, handle that with your normal chain of command. I don't know if the telephones are working, they probably aren't. It is your responsibility to get word to your officers. I will report to headquarters that all of you were present for duty here at the motor pool for today, so you are free until tomorrow. Any questions?" The men saluted and Darpheel returned the salute and started to turn and leave, but instead turned to the sergeant and shook his hand.

"Good work, Sergeant Quist, and to the men, too," Captain Darpheel said to the little man who had been his aide during their trip into the watery hell of the delta. The sergeant saluted briskly and Darpheel returned the salute. He then smiled, said, "I hope we can serve together again, under a little better circumstances." Then he turned and left.

Until he left the motor pool lot, Akim had not seen much damage, but as he walked out of the main gate toward the boulevard that ran from the airport downtown he saw the devastation.

The brick and concrete buildings which faced onto the boulevard were, for the most part, still standing. However, behind them on both sides of the road had been block after block of wooden structures, ranging from mere shanties to two and three story flats. All were ruined.

The pieces of wood and corrugated metal and the personal belongings from a hundred thousand homes had blown hither and yon about the capital city. In places the debris was so thick on the road that road graders had merely pushed a single lane path through it on the six lane boulevard.

The sun had only just come up, but already the streets were filled with people. Some were carrying piles of possessions or suitcases and walked with determination and purpose. Others simply wandered, near naked, with blank stares. Captain Akim Darpheel joined the column of people heading in toward the downtown area.

An occasional vehicle, usually an ambulance or emergency vehicle, would pass Darpheel, but it seemed as though very few private vehicles were being used in Dhaka .

Akim Darpheel had only gone a few blocks when he heard a sound which reminded him of his boyhood.

"Allahhu Akhbar!" a mu'azzin was calling out the call for morning prayers. Akim had become so accustomed to the electronic amplification of the mu'azzin's call from the loudspeakers in minarets in the big mosques of Dhaka, that to hear a mu'azzin use his own unamplified voice in the high pitched bellow to try and call the faithful to worship was nostalgic. It also answered the question of whether electricity was working yet.

The call had its effect. Although he was not prone to being overly religious, Akim knew that he must answer the call and join in prayers this morning. His voice would pray for the millions of souls whose earthly bodies now floated in the delta or were covered by the rubble and debris all about him.

After prayers Akim continued his walk downtown. There were no taxis anywhere.

At one of the big traffic circles on the boulevard a man was hawking an edition of the Dhaka Times. It was only four pages in length, but Akim assumed that even a mere four pages had been an achievement for the printing plant in the condition Dhaka was in.

It was from this paper that Darpheel learned that the eye of the typhoon had indeed passed just to the west of Dhaka. This meant that the highest winds had hit the capital squarely and this city of seven million people had fared no better than Chandipur and Narayanganj. And the rest of the country had been similarly hit with the double blows of typhoon fury followed by the run-off of the typhoon's floodwaters. The dead were estimated to number in the millions and the homeless in the tens of millions.

Darpheel continued his trek. At one point he had encountered a man selling bottles of soda pop and a few items of food from a storefront, one of very few open businesses. When he had enquired as to the price, he was given a figure five times the normal. But he had not eaten any breakfast and had shared the last of Sergeant Quist's canteen of water in the pre-dawn hours. He tried to bargain, but wound up paying the asking price for a warm orange soda and stale breakfast biscuit.

As he passed the Technical College for Men where he had attended four years before, Akim Darpheel saw an appalling sight. The gymnasium was obviously being used for a hospital as ambulances and teams of stretcher bearers were coming and going. But the soccer field had been turned into an open air mortuary.

Row upon row of bodies were lined up the entire length of the field. He could see from the sizes of the bodies and the colors of clothing that the rows were segregated by age and sex. Captain Darpheel estimated that as many as ten thousand bodies covered the soccer field. Half again that many people ambled in snaking queues amongst the dead, looking for loved ones. From time to time, one of the mourning searchers would kneel at the foot of one of the dead or collapse and hug the deceased. This was the signal for on of the white-smocked attendants to come with clipboard and toe tag to identify the body and order its removal from the field.

At this sight, Akim Darpheel gave thanks that none of his family lived in Dhaka. He could only pray that his family's home in the foothills of Assam fifty miles north of Dhaka had been spared. Indeed, he could only pray.

Akim Darpheel stared open-mouthed at his former home. He was not alone. A large crowd of on-lookers watched rescue workers sort through the rubble.

Captain Darpheel had lived in what was one of the few multi-story buildings in this area of Dhaka. Owned by the government, the second and third stories served as bachelor officer quarters for the military and national police staffs who worked in the nearby buildings which housed many of the national government functions of Bangladesh. The upper floors were rented as family quarters for the managerial personnel of Bangladesh's civil service. The top floor had the satellite and microwave antennas for the Bangladeshi national communication system. Being of sound concrete construction with concrete inner walls, the building had been designated a shelter and Akim could only imagine how many people had been in it during the storm. Now it lay, its floors like a fanned deck of playing cards, across the boulevard. The upper floor had hit the building across the street which had reciprocated by collapsing back over the first.

A Caterpillar tractor with a drag line was pulling the flat concrete chunks away and another pushed them to the side of the road. As Akim watched he could not tell if the primary purpose was to remove the bodies from the rubble or to clear the street. However, there was a team of soldiers who ran in after each slab was removed and dragged the bodies to an ever growing pile on the opposite side of the boulevard.

It was clear that only Darpheel's sojourn in the waters of the typhoon ravaged delta had saved him from death in his own flat.

Akim saw an army captain, who he knew, directing the crew of soldiers. He ran over and saluted. The haggard captain squinted at him through the dust and sweat in his eyes and then returned the salute.

Akim stumbled for words, "Do we know how it happened? How many? I have been with one of the search patrols, I did not know?"

The captain wiped his brow with a sleeve, "Yeh, yeh. I'll tell you what I know. They say that at the height of the winds the building just lifted up like a box kite, all the glass falling out. And then it just settled sideways across the street. Not many witnesses, everyone was under cover themselves. From what we have found since we started there were hundreds and hundreds people in the interior hallways. No survivors, it is pretty grizzly when maybe a thousand people are pancaked between tons of concrete slabs. We wouldn't be doing this yet, we don't have the right equipment, cranes for instance, but the Acting Prime Minister, General Kadiim, ordered the boulevard cleared so the dead and injured could be moved through the city. But, this is nothing compared to the rest. Did you hear about the west section?"

"No, what happened?" Akim could imagine nothing worse than the ghoulish sight before him.

"They say the storm itself, coming right up the main river channel, broke over the river levees and flattened almost all the houses and buildings. Then the rain and flooding hit early yesterday. Both rivers, either side, flooded meters deep and the whole section of the city washed away. Someone said a million people are missing or dead." The captain shook his head.

"Well, all I can say is that the same happened at Chandipur, the whole city is gone," Akim replied. The two men looked in each others eyes for a moment.

"May Allah be with you."

"And his peace unto you, my brother."

The captain returned to his men who were just clearing the next layer of concrete off of the pile.

Akim looked for a minute at the second layer of concrete to his left, which represented the resting place of all of his comrades and roommates who had not gone to the countryside. He mouthed another prayer of thanks that his normal headquarters function of training administration had not been a critical task that would have kept him in the capital instead of on the disaster response team. He also noted that his entire worldly possessions now consisted of one dirty Army uniform.

Akim carefully picked his way across the rubble blocking the boulevard. He still had to get to headquarters some three blocks ahead.

Akim passed the toppled building and had a good view of the central governmental complex. Most of the buildings were intact, including the general staff headquarters. But, as he approached the center of the city, he saw that the huge plaza between the Parliament Building and the general staff and other buildings was filled with tens of thousands, perhaps even a million, people.

As Captain Darpheel came closer, the crowd took on the appearance of a churning mass of bodies without organization. They were a wretched looking bunch. Many were injured. Some, seeing his military uniform, came over and pleaded for assistance in finding a loved one or food or medical care. He could only shake his head and shoo them away. But, the attention to his status did cause him to straighten his uniform and walk more proudly. He rubbed his face in his hands to insure that the grime of the journey did not show. Only his position as a military officer separated him from these thousands and millions of his countrymen who were destitute. He hoped.

When he had come close enough to fully view the Parliament building area, he saw the reason for the crowd. The Army had set up a field hospital and kitchen on the grass. There were soldiers with fixed bayonets ringing them both and keeping the crowd in order.

He had forgotten to ask the captain why the General was Acting Prime Minister. What had happened to the Prime Minister? Had she died in the storm or had the military finally carried through on the threat it had made during the last such disaster and taken over the government? He would find out as soon as he got to headquarters.

———

March - Los Angeles

The silver gray Mercedes accelerated around the merge ramp from the 710 to the 10 freeway. Ian Petersen kept his car to the right since the exit to downtown Los Angeles was just ahead. The big sedan's air conditioner labored to keep the car cool, but there was nothing it could do about air quality. The misty burning of the eyes, so common to Los Angeles, had hit him.

He took the exit and made his way through the bumper to bumper traffic of downtown L.A. to the Arco building where he usually parked on these pilgrimages to his mortgage banker. He had learned early on that the only exit from the neighboring bank building's parking garage and its surrounding construction forced him into a nearly impossible-to-escape-from route out of the city, the wrong way. He wondered why no one ever tried to coordinate such things as construction, so that you did not have two different construction crews closing a vital street like Figueroa off from either sidewalk for two lanes out, leaving only one or two lanes for a torrent of cars to squeeze through. Like many people who were not native Angelinos, Ian had a suspicion that it was a myth that anyone was really in charge of this sprawling behemoth of a city. Everyone just did what seemed OK for the moment and pretended the government had control of things.

Luckily, this would be his last trip downtown for some time. Today was closing day. His assumption of the mortgage on the apartment complex was to be complete. If all went well, between his restaurant and his new apartments, he would not have to leave his beloved Newport Beach for a long time. Things were not so crazy in Newport. His ten years there had taught him the intricacies of everything that was important. Society, money, government, food, cars, fun, sex... everything.

He had first come to Southern California as the front man for the Canadian real estate syndicate that had taught him the ropes. As a natural deal maker, he had flourished in the freewheeling market of the California coastal communities. Within a few years he had gone independent and made a tidy sum for himself. Now with the acquisition of a trendy harbor-side restaurant, a ready-made family of his beautiful new wife and her two children from a previous marriage, he had just about everything he could hope for. And in an hour or two he would have a seventy unit apartment complex on the edge of Newport Harbor that was just about ready to go condominium. His fortune was made. Pretty good for a farm boy from Ontario.

The striped wooden arm blocked his way into the parking garage. Ian pushed the button to roll down the window and he was met with a rush of hot air that made the skin on his face tighten. He grabbed the parking stub and rolled the window up.

It had been warm in Newport, but he had forgotten about the early heat wave that had been forecast until it rolled in through his window. That explained the bad smog, too. Hell, it was still March and L.A. was already becoming a tinder box. Thank god for air conditioners, and living at the beach.

Ian kept his coat off for the walk up the ramp and across to the Arco building. By the time he reached the tinted glass entrance he had worked up a moderate sweat. As he crossed the lobby to the elevators he saw the broker who had handled the sale for the seller waiting for an elevator. The fat little real estate man was mopping his brow with a rumpled kerchief. As he came near, Ian could see dark circles of sweat bands staining the suit jacket under Bromley's armpits.

"Hot enough for you?" Ian said cheerfully.

"God, do you believe this?" J.J. Bromley replied. "It's only five blocks from our office. I thought I'd walk. Bad move."

"There ought to be a law against heat waves before Easter."

The elevator arrived. Bromley pushed tenth floor. Ian looked at him through furled brows. Ian said," It's twelfth floor, unless you know something I don't."

J.J. Bromley looked puzzled, "Yeh, that's right. Guess I'm just not thinking." He pushed twelve.

The closing went like clockwork. The escrow lady had everything in order for the loan, Ian had the down payment in certified funds and J.J.'s corporate client had the deeds in order.

"Congratulations, Mr. Petersen, I think the number of zeroes on that mortgage qualifies you as an honest to goodness real estate mogul." J.J. said as he folded his commission check into his pocket.

"I think I will wait for mogul status until I get the mortgage to shrink a little. It's still a bit scary." Ian was only half joking. He had inked numerous big deals for others, but this was the first eight figure deal he had ever put his own name on or at least his own corporation's name. The purchase money loan he had needed from his former employers for the down payment had been the first seven figure deal. It was a bit heady, even for a man who aspired to high places.

"Did you bring over the original architects drawings you promised," Ian asked as he and J.J. prepared to leave.

"Damn, I'm sorry Ian. I had them right at the office. I'll get them out to you by courier this afternoon."

"No problem, I'll just walk over with you and get them now, if that's all right."

"If you want to, but I really can send them out."

"Naw, I'd rather have it over with. I want to get them to the marketing firm I hired to bring the tenancy rate up and make plans for the condo conversion. Got to do something to fill that place and pay the bills."

They went down the elevator and back out into the noonday heat. Both men took their coats off for the walk. They joined hundreds of office workers on the crowded sidewalks.

Neither man tried to make conversation considering the crowds and the heat. They just walked and waited for crossing lights.

In the middle of the second block, Ian noticed J.J. falling back from Ian's pace. Ian slowed and fell back to him. The little man was openly panting.

"A little out of shape. Nothing thirty pounds wouldn't cure." J.J. smiled meekly as Ian met his gaze.

"You want to stop and rest." Ian didn't like the look of J.J.'s pouring sweat and labored breathing.

"It's just another block." J.J. trudged on.

At the next corner J.J. leaned on the crossing light pole facing to the street away from Ian. From behind Ian could see the little man's shoulders rise with each quick breath.

The "Don't Walk" light changed and J.J. did not move. Ian moved around him and looked into his face. His skin was almost gray, and his eyes were unfocused. As Ian watched J.J. Bromley' grip slid down the pole and, but for Ian's grasp, he would have fallen into the gutter of the intersection.

Ian pulled J.J. onto the sidewalk. Bromley's face was, is spite of the heat, cool to Ian's touch and his brow now had much less perspiration than Ian's own. Ian started to try and find a pulse, but a passing physician stopped and took over the first aid duties. An ambulance was called, but it unfortunately took the route through the Figueroa Street blockade.

———

Ian had barely known J.J. Bromley, and Ian's wife, Cyndi, had never met him. But it seemed appropriate that they go to the funeral. The little real estate man had helped put together the deal that would make Ian and his wife independently wealthy. The heat wave continued and the assembled mourners at Forest Lawn Cemetery partook of a bit of J.J.'s final agony, black formal clothing made very poor hot weather clothing.

A statistician at the Department of Health made note of the cause of death on J.J. Bromley's death certificate. It was the first case of death from heat stroke for the year.

———

April - New London, Connecticut

VIA NAVCOMMFAC - 2 PAGES XMITED
UNCLASSIFIED - EFTO - FOR OFFICIAL USE
ONLY
FM: CMDR COHEN, USCG, NEW LONDON, CT
TO: GAIA OFFICE, NCAR, BOULDER, CO
INFO: SCRIPPS INST(ATTN: DR. COLTON), SAN
DIEGO, CA
GAIA DISTRIBUTION LIST DELTA
SUBJ: SITREP OF SIGNIFICANT MARITIME
INFORMATION - MARCH
1. USCG AND INT'L SOURCES REPORT
SEVENTEEN PERCENT INCREASE IN CONFIRMED
ICEBERG SIGHTINGS OVER NORM FOR FEBRUARY-
MARCH TIME PERIOD IN NORTH ATLANTIC.
SIMILAR REPORTS FOR SOUTHERN HEMISPHERE
FOR SIX MONTHS PERIOD.
2. FROM COLLATERAL SOURCE (NAVY): McMURDO
STATION, ANTARCTICA, REPORTED LONGEST AND
HOTTEST SUMMER PERIOD ON RECORD. HIGHEST
TEMPERATURE EVER RECORDED FOR ANTARCTIC
REGION RECORDED AT VANDA STATION ON
JANUARY 6TH. TEMPERATURE REACHED SIXTY-
FOUR DEGREES FAHRENHEIT. SEVERAL RESEARCH
AND RESUPPLY FLIGHTS DELAYED THIS
SUMMER SEASON (DEC-FEB) DUE TO SLUSH ON
NORMALLY SOLID ICE RUNWAYS IN INTERIOR
OF ANTARCTICA. RECOMMEND FULL INQUIRY VIA
SCRIPPS, NOAA AND PENTAGON.
3. USCG DATA FOR FIRST QUARTER SHOWS
TWENTY-ONE EASTERN SEABOARD REPORTING
STATIONS HAD HIGHEST EVER HIGH TIDE

LEVELS. ONLY TEN OF THESE HAD ANOMALOUS
WEATHER PATTERNS TO ACCOUNT FOR HIGH
TIDES. DATA FOR WEST AND GULF COASTS
AVAILABLE NEXT WEEK.
4. USCG CUTTER (USCGS CAPE HATTERAS)
ON MARCH 8TH REPORTED MASSIVE ICE FLOE
ADRIFT 147 NAUTICAL MILE EAST-SOUTH-
EAST OF CAPE FAREWELL, GREENLAND. SIZE
ESTIMATE WAS THIRTEEN BY EIGHT NAUTICAL
MILES. ICE WAS FRESH PACK ICE, NOT SEA
ICE. PROBABLE SOURCE - KING FREDERICK
GLACIER FIELD (62.00N 43.50W). THIS
PHENOMENON IS SIGNIFICANT FOR SIZE,
COMPOSITION AND TIMING.
XXX

———

April - Washington, D.C.

The White House guard was a serious-looking black man in his fifties. He nodded acknowledgement to the well-known Senator. He had already memorized the expected appointments that would be passing through the executive entrance to the White House on his shift, but still checked the clipboard out of habit and duty. He eyed the Senator's big briefcase, but declined to ask the former presidential candidate to open it for inspection. The metal detector the Secret Service concealed in the entranceway would suffice. He pointed the Senator up the path to the double glass doors. Near the doors a glowering man in a blue raincoat stood watch. It had not rained in weeks. Andy knew that suspended beneath the agent's armpit under the raincoat would be a MAC-10 or similar weapon.

The Secret Service agent was a young man, not one of the many Andy had come to know during his presidential candidate status years before. The agent opened the door for Andy, and said, "You can wait in the third room on the right."

Andy had been to the White House many times before, but usually to the press and meeting rooms or through the formal entrance for a state occasion in the East Room or the State Dining Room. He now entered the West Wing of the White House and on the far right corner of the wing the Oval Office was located. This was the route used primarily for visiting the President or the Executive Offices, the regular White House business was handled through the business entrances to his left or on the East Wing.

As Andy turned into the third room on the right he happened to look back and the agent was looking in through the glass-door to make sure the Senator did go into the right room. Andy smiled to himself.

The third room on the right was a well-appointed waiting room. To one side there was a cloak room and to the rear were two doors marked "Ladies" and "Gentlemen." Andy took a seat on a brocade couch setting his briefcase on the floor.

The White House seemed remarkably quiet, much more so than on Andy's previous visits. However, this was a Saturday morning.

Andy did not have long to wait. Very shortly, the Vice-President came down from the other end of the hall and greeted Andy.

"Good morning, Andy. The President's waiting and we have almost an hour clear depending on when the next appointment arrives." They headed up the hall together. Andy was taller than the Vice-President, but had to walk quickly to keep pace with her. When they spoke, Andy's throaty voice and New England accent was distinct from the California accent of the Vice-President.

"I really appreciate your doing this."

"Not much else I could do, after the sales job you gave me Wednesday evening. Pretty frightening. By the way, I had to let him know something of what you're bringing to him to get him to open a spot for us. He's delaying the trip out to Camp David in order to meet with the Bangladesh ambassador anyway, so we snuck in beforehand."

They passed a young woman standing near an alcove in the hall. She appeared to be a secretary, attractive with a short bob haircut. But, Andy caught a glimpse of the wire running under her hair from her ear, down her neck and into her clothing. Andy knew that somewhere in her sleeve would be a matching wire to her radio microphone. The Secret Service kept each member of their executive protection teams in constant contact with one another. The Vice-President opened a door and she and Andy entered the President's anteroom.

A middle-aged woman behind a desk rose and said, "He's waiting for you, go on in," as she opened the door into the Oval Office. An older man sat in the corner of the room reading a magazine. His earpiece was more obvious than the woman's in the hall had been.

The President was standing in shirtsleeves looking out the arc of windows behind his desk when they entered. He came over and shook Andy's hand.

"Andy, it's been a long time. I think the last time you and I had time to sit down for a talk was at the economic summit during the transition. We should keep in better touch."

The President motioned Andy and the Vice-President over to the chairs and couch in the center of the office. "Let's have a seat."

The President took a chair at the end of the low coffee table and Andy sat on a couch with the Vice-President sitting opposite him in another couch.

"Well, Senator, what's up. The Vice-President tells me you have something big up your sleeve and want to give me a look."

"I guess you could say that. First, I want to thank you for meeting me. I think you'll see why I thought this important enough to handle this this way."

"If both of you think this is important I have no doubt that I will too." The President shifted back in his chair waiting for Andy to speak.

"As the man who gave me this information last week said to me, it is hard to find a place to start. Are you familiar with the GAIA project?"

"Not really. Some sort of weather satellite named after a Greek Goddess isn't it?"

Andy withheld showing his amazement at the President's lack of knowledge of the billion dollar federal program, but the President would not be alone. It was his first indication that perhaps the upper echelons had not been kept apprised of the important information Brady had tried to send up. He continued, "Well, yes, in part, it does involve several weather and geophysical survey satellites, but the actual GAIA project is a function of the National Oceanographic and Atmospheric Administration in their Office of Climate and Research. The name Gaia does come originally from the Greek goddess of the earth and life. A British scientist by the name of Lovelock in the early '70's used the name Gaia to describe his theory that the whole planet is really a self-contained balanced ecological system, in reality a living thing, in and of itself. The theory has gained pretty wide acceptance over the years. To those in the environmental movement the name Gaia has come to symbolize the balance of life on Earth.

"Our government project, GAIA, stands for Geophysical Atmospheric Interactive Analysis and it is a major project that integrates information from hundreds of different sources and tries to come up with a real picture of the environmental outlook for the Earth. For years we've had all kind of theories about "greenhouse" this and "global warming" that, but nothing that could give us a true picture of what was really happening or what we could expect to happen if we took a certain course of action. The project is exactly the kind of thing you had in mind when you spoke to the Earth Day rally out at the University of California and promised that your administration would set a priority of creating an interdisciplinary approach to studying the problems of the environment and global warming. Andy knew as soon as he said the words that his background research was a bit to transparent.

"You're not going throw old campaign promises at me, are you?" The President said, chuckling.

"Not at all, especially when the program is up and running. Actually, the GAIA program turned out just perfectly. It was conceived at N.O.A.A. several years ago. I got involved when it came through my sub-committee on the hill as did the Vice-President. It was set up about two years ago and they have the best people, dozens of the finest scientists, the best equipment and the system works perfectly."

"So what's the big problem?" the President asked the obvious.

"The problem is that they have the answers that we have spent billions of dollars to get, those answers are of earth-shaking importance and the people at GAIA can't get anyone in your administration to listen."

The President scowled and simply said, "Explain!"

Andy drew a breath and continued, "I have in this briefcase, information from the GAIA project which forecasts truly catastrophic impacts on our country and the world, with a very high degree of probability, for both the short term and the long term. And it appears that members of your administration, possibly with the best intentions, but out of a happenstance of political zealotry, loyalty to what they believe is your political and economic agenda and ignorance of the importance of what GAIA is telling them, have tried to bury in the bureaucracy what is possibly the most important information you or any other president has ever had to deal with."

The President sat forward in his chair, his eyes piercing into Andy's, "I suppose you can prove such a statement."

It was Senator Knowles turn to stare down the President. "Mr. President, my presumption can be proven by you. If your Secretary of Commerce has informed you that your government's top experts, who work for him, have reported a better than even chance that the cities of Miami, New Orleans and Houston, among others, will disappear from the face of the Earth in the next thirty years, then I am wrong."

The President's only reaction was to tense his jaw muscles. He looked at Andy silently for a moment, then over to the Vice-President. Finally he turned to Andy, pointed to the briefcase, "Let's see what you've got."

Andy shifted in his seat and began.

"As I understand it, the meat of the problem is the global warming we've all been hearing about and speculating about for years, but the problem is not as simple as that. What GAIA does is take data from hundreds of sources, from international weather reports, historical data, from satellites, from oceanography studies, from forestry information, acid rain, pollution, population statistics, more things than I can name; and they have created a computer system that can analyze all of this data and come up with what should be the most probable outcome of any giving set of facts. So this is a lot more than just one group of scientists looking at a graph of temperatures and telling us its getting hotter on old planet Earth.

"Actually what they have is a network of experts and computers that each handle one area of information, say climate or ocean currents, and then that information is fed to the big supercomputer at the N.O.A.A. site in Boulder and the GAIA computer puts together everything and comes up with a model of how the whole system will react to the data it receives. The real power of the idea is that then the big GAIA computer can feed the results back out to the field data centers and they can re-interpret how their particular specialty will react to the new model the GAIA computer has come up with, that's called interactive distributed processing. Apparently that is what makes this so much different from anything else, the fact that hundreds of variables can be analyzed at once and factors around the world can be tested for their effect on the whole earth environment. Here is a chart they gave me showing the various constituent parts of the GAIA team. As you can see everyone from NASA to the military to the Interior and Energy departments and several foreign countries plays a part in imputing the data they use." He handed the chart to the President.

"What they have come up with at GAIA is the idea that these hundreds of factors all at work on the weather, the oceans, the polar icecaps, and the forests have created a catalytic effect, or perhaps the better word is synergistic effect, on the earth's climate that is just now coming to a head. The idea that all of the factors involved in the earth's climate can build up for a long time and this climate system will struggle to remain stable for years and then it reaches a point of no return and just flips over into a new and decidedly different phase was brought out way back in 1987 at one of our Senate hearings by a scientist named Broecker. This drastic flip theory has been supported in large part by what GAIA has taught the scientists. They say that some things have gone beyond the point of no return. There are some things we can still do something about. And, lastly, there are some drastic steps we can take to try to restore the natural balance of things. But the bottom line they give us is that the mess we have created is changing our world in very drastic ways.

"If we continue with life as usual, they say that this will be a map of the United States in the year 2050." Andy handed the President a chart that showed a current map of the United States with blue ocean-colored shading covering much of the Eastern coastline, almost all of Florida, most of the Gulf Coast and huge bites of California and the Pacific Northwest. "That is a result of just a part of the polar icecaps melting coupled with the thermal expansion of the oceans. It could be worse. And the limited melting of the icecaps and the rise from thermal expansion have already begun, you've heard of the monster tides, the Boston airport incident, there are many examples.

"Perhaps even worse than the coastal inundation, if you can get worse than that, is the effect that the added heat will have on the rest of the climate. The GAIA computer has given a prediction that if nothing is done to stop it, the weather will change so drastically that it will hardly resemble what we now know." Andy handed the President another chart with two world maps and color bands showing arctic, temperate, forest, desert, subtropical and tropical zones for present and fifty years ahead. The President studied it and handed it to the Vice-President. "The food production patterns and population centers of virtually every country will be affected. Today's breadbaskets may be tomorrow's desert wastes and today's deserts may be coastal wetlands."

Andy continued, "Now, as I said, those are worst cases if we keep doing everything as we are now even with all of the measures we have tried to implement and all of the cards go against us. But GAIA also has that interactive processing capability that accounts for changes. And, sorry to say, if Boston, New York, Philadelphia, Houston, Los Angeles, London, Amsterdam, St. Petersburg, Tokyo, and the rest of the world's seaport cities are all under water and a billion and a half people are homeless, the fact is that pollution will be less and other factors will change so the actual situation will be slightly less than these scenarios. GAIA can tell us that if we want."

Andy stopped and handed the President two more charts before reaching for the water pitcher and a glass from the table. He took a drink and continued.

"Now, Mr. President, when they gave me this information I asked the same questions you are probably thinking about. What are the chances that any of this will occur and the chances that this is all a big computer error or just more of the same speculation we have all heard about for a couple of decades now? They answered that they had checked into that and came up with probabilities of the best case and the worst case and then came up with a very highly probable 'most likely case'. Then they gave me this." He handed an inch thick report to the President.

"These are reports of the various unique phenomena that are occurring in the world around us, its the facts that keep these reports from being mere speculation. You are meeting this morning with the Bangladeshi Ambassador. Their recent cyclone had the highest sustained wind speeds ever clocked on earth and the highest tidal surge ever seen. And cyclones and hurricanes are fed by the earth's heat and the evaporation of ocean waters. Twelve of the worst hurricanes and cyclonic storms in history have occurred in the last ten years. My home got hit by Hurricane Sandy and you toured that damage yourself when you were V.P. When Hurricane Ida hit there was a lot of talk about it being caused by global warming, but this is even more than that. The U.S. Coast Guard reports that virtually every tide table for ports in the U.S. has had to have its highest high tide marks changed upwards in the last two years. Some by several feet. The U.S. Park Service is reporting serious unprecedented beach erosion on national seashores and coastal wetlands. I could go on and on. It is well-established that global warming is increasing hurricane threat. It is all right there."

"And then I asked them what the best case scenario is, if we pay attention to this and do everything we, meaning the whole world, can do to prevent this and we get lucky. And even that is enough to scare the bejeesus out of you. The Coast Guard officer on the GAIA staff said the best case is that we will have a sea level rise of between two and seven feet in the next twenty to fifty years. The meteorologist said we can expect repeated instances of killer hurricanes, and worldwide, global warming on the average of two to five degrees in the next decade. And mind you, that average is moderated by cooler climates and huge areas of relatively stable ocean climates, so the hottest areas could have temperatures and prolonged heat waves beyond anything we have ever encountered on earth. Likewise, within that average temperature rise could be just enough warmth for places we know as cold to warm up considerably.

"The agricultural specialist on the GAIA staff predicted, remember this is best case, massive and repeated crop failures in the world's primary food producing areas, including the American Midwest, Russia and the Ukraine, and China's prime rice producing areas." Andy stopped for a moment and let the President think.

Andy picked up another stack of reports and charts and set them in front of the President. "What I just gave you was the best and worst case. This pile here is the probable outcome. Assuming the law of averages and the whims of politics take a middle ground and assuming that the leaders and people of the world pay a modicum of attention to this and do a reasonable amount of work to prevent it. It gives us a picture of an America in not too many decades where 14 % of the people have been relocated by coastal flooding. Where gross national product and our agricultural production is diminished by ten to thirty percent, maybe more. And I suppose it is fitting that you have an appointment this morning with a diplomat from Bangladesh, because that is one of the countries that is on the GAIA list of nations and areas that will cease to exist for all intents, along with Denmark, the Netherlands, as well as much of Egypt, the Philippines, Indonesia, and even Great Britain will lose...."

There was a knock of the door and a moment later the president's secretary opened the door to say, "Mr. President, the Ambassador is inbound and will be arriving in ten minutes." She quietly closed the door.

The President thought for a moment and then said, "I have to do this thing with Ambassador Rahman, but I want to continue this if possible. How are you set for the rest of the morning, Andy?

"I'm at your disposal."

The President turned to the Vice-President, "How about you?"

"I can get free also."

"Good, let's get this in to the Cabinet room," indicating Andy's charts and reports, "I'll be in as soon as I can break free from the ambassador."

The President walked over to the door and opened it, calling to the secretary, "Gloria?"

Gloria came in, followed by another woman that Andy recognized as the White House Chief of Protocol. The President was putting on his suit coat behind his desk.

The President spoke to his secretary.

"I'm canceling the Camp David trip, something has come up. Tell my wife, I don't think she really wanted to go anyway. Let everyone else know. Also, I saw Admiral Kirksey around here earlier this morning. If he is still here I'd like him in the Cabinet Room. If not, give him a call to come in, and call Doctor Fairchild in to." Kirksey and Fairchild were the President's national security advisor and science and technology advisor.

"Should I tell the Chief of Staff?"

At the mention of the Chief of Staff the President looked up and over to Andy and the Vice-President. They said nothing. Thinking for a moment, the President finally said, "No, he's going to his wife's speech in New York, let him relax for the weekend. Of course, let his office here know we've canceled Camp David, but don't bother him and tell his staff not to. Kirksey and Fairchild are who I need to hear from. And if you can't get them, get their deputies in here, maybe both them and their deputies."

The President left with his Chief of Protocol followed by the secretary who closed the door behind her, leaving Andy and the Vice-President alone in the Oval Office.

The Vice-President picked up several of Andy's charts and reports and headed to the door on the opposite side of the office.

"I'd say your mission was accomplished. He seems to have got your message."

Andy put the remainder of his reports into his briefcase and followed the Vice-President.

"I am afraid we all know the mission is just beginning."

———

April - Lake Ladoga, Russia

Commander Litasku buttoned his collar flap on the brown greatcoat closed to cut out the wind. He wished he had his winter cap with the ermine earflaps, instead of this useless round military "saucer cap." The driver had rammed the throttle forward and as the boat picked up speed the chilly wind on the flying bridge increased in intensity.

The St. Petersburg militsia had inherited three patrol craft from the old KGB Territorial Guards. In fact, the coxswain who drove the boat was one of the men who had transferred to the regular police after most of the internal KGB functions had been disbanded. Other than the inspection when he took charge of the craft, Istvan had only had occasion once before to even board the boats; to investigate a murder in a cargo vessel at anchorage near St. Petersburg. Now he was glad he had them under his command. The three fifteen meter steel hulled craft were well-equipped and sturdy, one even had a reinforced prow, to make it a miniature ice breaker. They had been made in a Latvian shipyard while the Latvians still did business with the KGB, rather than the American and French yacht merchants who now controlled the shipyards in Riga.

He had heard reports of the horrible flooding problems to the east of the city and he wanted to view them first hand. The high tides had continued in the port and a stationary weather system had continued the inclemently warm weather. Most of northwestern Russia drained into the river system that eventually led to Lake Ladoga and the Neva River up which he was now cruising. Ice and snow which normally did not melt until April or May and sometimes even June had melted in early spring, taxing the river system. The problem was compounded by the fact that the farmers were not yet ready to plant their crops and use the water in the fields, so even that water which normally would have been used for irrigation was going downstream unused.

Technically, Lake Ladoga was not in Litasku's police region, but he was intent on knowing as much as he could of the flooding which had immobilized his city, flooding subways, roads and buildings. His colleague who had responsibility for Ladoga had his hands full with the flooding upstream of the huge lake and certainly would not mind Litasku's venture into his turf.

As they proceeded up the river they passed a series of barges which were used to transport coal, wheat, pig iron and any other bulk product to be transported throughout the river and canal system that crossed northern Russia. The same glacially levelled terrain that made canal building possible was what made flooding so devastating. Once flooding started there was no place for it to run off to.

"Quite a mess, no?" Litasku said to the coxswain. The man could only shake his head in disbelief at the acres of flooded farms on either side of the channel.

They had left the last of the suburban highrises a few miles back. Now, only farms and small villages could be seen. Most of the low-lying area was under water. To the south some of the land was slightly higher and there were a few wooded areas. To the north, toward Ladoga, it was all water. Their view to the south was cut off by the embankment of the main East-West railroad line which edged closer to the river as they went farther East, nearing the entrance to Lake Ladoga.

Istvan could now see a work crew on the railroad embankment. They had a railroad car crane, a flatbed and a work caboose along with a small yard locomotive. About thirty men were gathered at the site. The police boat had been on the north side of the river, now Istvan ordered the coxswain to cross over to look at the railroad work.

As they came closer Istvan could see that the men worked in haste. Most were filling sandbags on the flatbed and carrying them to the other side of the embankment, out of the boat's view. One man, obviously the crew foreman, stood on the landing of the caboose shouting orders into a megaphone.

The coxswain idled the boat's engine at Istvan's orders, just maintaining their way in the river current. Istvan could now see the problem. The near side of the embankment was scared by the marks of erosion. The gullies in the brown-green sod of the embankment and two missing railroad ties showed that the water on the other side must have come over the top of the embankment, and recently.

The top of the embankment was at least a dozen feet above the level of the river and lake. Istvan wondered how big an area was flooded behind the embankment out of his view.

The crane's diesel engine roared to life and workmen loaded a long piece of steel I-beam into the crane's cargo loop. The I-beam was set carefully in place on the other side of the bank. The loop was released and men scurried to cover it up with more sandbags.

The coxswain nudged Istvan with his elbow and pointed. Following his finger Istvan could make out a spot in one of the lower gullies on the near side of the embankment which had started to gush muddy water. Everyone on the embankment was so intent in working on the far side they were not paying attention to this danger.

Seeing the potential for disaster, Istvan tried to shout to the foreman. He was too far away.

"Hit the air horn." Istvan ordered the coxswain.

Even though he expected it, the booming honk of the air horn startled Istvan. The whole railroad crew turned their attention to the police boat now.

Istvan motioned to the foreman and pointed to rupture inn the embankment. "A break! A break!" he shouted.

The foreman looked in wonder at what the militsia officer was trying to do, but then he got the message and saw the flow from the embankment. He shouted to the crew, something Istvan could not make out. The men who had been placing the sandbags carefully near the I-beam now started to grab the bags with a man on each end and swing them back and forth trying to throw the sand bags down the far side of the embankment to where the tunnel breach in the embankment was, apparently underwater.

The gush from the gully continued unabated. The foreman and one other man climbed down the near side of the bank to view the break close up. They slid much of the way down in thick mud.

Just as they reached the area of the gully a crescent shaped chunk of earth above the hole broke away and widened the breach. Then another, larger chunk from the side of the hole broke out and the man with the foreman had to jump to get clear of the torrent now opened. He was able to get away from the rush of water, but his movement sent him sliding to the bottom of the bank, into the river.

The foreman hurried to the top and shouted more orders, waving his arms to the men and the engineer in the locomotive. The engine belched sooty smoke and started pulling on the work train, slowly inching forward. Men threw shovels and equipment on the flatbed and jumped on board themselves. The locomotive and the crane were now forward of the hole in the bank, the flatbed directly over it and the caboose behind it.

The size of the hole now increased each second, with more and more muddy water and soil gushing out. The man who had lost his footing and fallen in the water now gave up trying to climb onto the bank past the ever widening breach. Instead, he turned and swam upstream away from the danger.

The force of the escaping water now tore huge chunks out of the bank. Soon the breach reached the top of the bank and the ties and track no longer had soil beneath them. The train had started to roll now, with the flatbed clearing the new-formed chasm.

The caboose was just crossing the chasm when a large chunk of soil fell away from the forward side of the hole. The caboose quivered for a moment and then the ties parted from the rails and there was nothing more to hold the caboose. The rear pairs of wheels on the caboose slipped between the tracks and the aft section of the caboose tilted into the torrent.

One person from the caboose, apparently a young woman, jumped from the caboose to the flatbed. Two men on the flatbed grabbed her to keep her from falling off.

The foreman was now peering over the rear end of the flatbed, surveying the expanding chasm that was sucking his caboose in. With a sledgehammer he hit the release bar on the caboose hitch which now held the rest of his train in danger. After three strikes at the bar the hitch let go and the train lurched forward.

Just as the hitch let loose with the full weight of the caboose on the unsupported rails, a weld in the nearest rail broke and the caboose flopped sideways into the torrent which was now twice as wide as the railroad car. The caboose bobbed and spun in the flow of water from behind the bank. It floated about twenty meters out into the river and sank, gushing brown water bursting through its top-side windows as it went down.

The coxswain had been backing the police boat back out of the immediate area of the break. They were now about thirty meters out. The water from behind the bank still continued, as it continued to wash out more and more of the main rail line that connected St. Petersburg to Moscow and the rest of Russia.

Commander Litasku yelled down to the young lieutenant who normally commanded the boat and who now stood agape at the deck railing below, "Radio headquarters, and let them know what happened. Make sure they tell the railroad switchyard." After a moment he added, "And have them contact Moscow and Novgorod militsia headquarters."

The young officer nodded and saluted before disappearing into the cabin.

Istvan Litasku stood shaking his head at the sight. The one lone man in the water had clambered his way back up the bank , but was on the tracks hurrying away from the widening channel through the tracks. The work train had retreated a respectful distance to the west and they were watching the spectacle like he was.

Someone should have told the authorities that a huge backup of water had formed behind the rail embankment. Something could have been done in time. Now, the massive surge of water had destroyed the railroad, had rushed into the Neva all at once and would be flooding more of St. Petersburg in hours.

Commander Litasku sighed and told the coxswain, "*Domoi.*" It was time to go home. There would be more problems to deal with back in the city.

———

April - Washington, D.C.

Dr. Fairchild and his deputy were arranging a series of graphs at the end of the big Cabinet Room table. Admiral Will Kirksey was still looking at the coastal inundation maps. Both Senator Knowles and the Vice-President were seated in the middle of the table opposite of where the President had been sitting before he had gotten up and moved to the window, where he now stood, looking out.

"Doctor Fairchild, we've all heard the Senator out. Now, I know there is a lot of stuff to sift through here, but what is your first blush reaction to all this?" The President continued to look out the window rather than to Fairchild as he spoke.

Fairchild straightened up and turned to the President, he still held one of the graphs. "I've always been excited about what GAIA could do for us. It's just that I was totally unaware that they had put everything together to give this kind of result. I was involved in the first symposiums to decide what the organization of the GAIA network would be, but I lost track of it when it when N.O.A.A. went into the actual technical design and implementation work. I must say that I am surprised, no, shocked, at the fact that this was not sent immediately to the Council on Environmental Quality. The Council is your advisory body on this sort of thing and I can't imagine this sort of thing being kept from it. Perhaps it is our problem for not keeping tabs on N.O.A.A. and Commerce, but it is clear that this should have been to us some time ago."

"You really feel this is solid enough to base our decisions on? Why should we give this study any more credence than the dozen others that have come in that say other things, or the naysayers that say the current scare about global warming is just an overreaction to a recurring phenomenon like the Dust Bowl in the thirties?" The President now turned to Fairchild.

Fairchild thought for a moment and then spoke. "Mr. President, when a man goes into a hospital with a heart problem, the first thing the nurse does is to take his temperature and then his blood pressure, but before he is out of there they have poked and prodded, and sampled and tested, and x-rayed and cat-scanned everything about him. And then if they think there is a problem they can call in the experts to analyze everything they have found. And the net result is a thorough medical report that gives not only an evaluation of the patients current condition , but a prognosis of his future health.

"Now, if you will allow me to be overly simplistic about this, the reports of global warming can be compared to the initial temperature and blood pressure. They each give an indication of a problem situation. And the final GAIA results we have here can be compared to the entire medical record. It gives us a clear picture of what is wrong and what we can expect. And an expert looking at either the medical report or the GAIA reports can look at them and give us a pretty good idea of what the outcome of each will be if we follow doctor's orders or if we don't."

"Then you really believe in this?" the President looked directly at his science advisor.

Fairchild spoke slowly, "I am sorry to say, I do. And it scares the hell out of me." Fairchild's deputy nodded his head to emphasize his boss' words.

The President turned to his national security advisor at the other end of the room. "Will, what do you think about this?"

The retired admiral looked up from the maps, "I think this is a real powder keg. Not just for what it mean if it happens, but for what it can do when we announce it or when we start to take action to avert it. Hell," he pointed to one chart, "if these are right, you have most of the world in chaos. When you announce that the United States government is expecting to lose every seaport and a sizable chunk of its infrastructure to ocean flooding and climate disasters, you will have economic repercussions of cataclysmic proportions."

The President cut in, "That right there is probably why this sat over at the Commerce Department until they could make sure, or whatever the hell they were going to do with this," the President was angry and turned back to his view out the window. "How in the world did we come up with those bankers and businessmen at Commerce running the weather bureau, any how?"

The President pounded a clenched fist on the window frame, "But, dammit, they should have let me in on it, whatever they felt about this. I don't see how they can equate a fragile economy with the need to cover something like this up."

The President touched the intercom box on the credenza between the windows.

"Gloria?"

"No, this is Margie, Gloria went to lunch."

The President looked at his watch, "Right, I want someone who can take steno in the Cabinet Room. Also have whoever is the duty officer in the chief of staff's office come in here. And have the kitchen send us lunch for seven. Oh, and clear my schedule for early afternoon."

The President resumed his seat at the big chair in the center of the table. He put his glasses back on and looked from Admiral Kirksey to Andy and the Vice-President to Fairchild and his deputy.

"Now, I would like to hear what you think we ought to do from here. I want to know what you think we need to do to clarify exactly what is facing us. What we need to do to organize a response. Who we should tell about this and what. You all know the kind of advice I'm looking for. Right?"

———

Everyone was still seated as they had been when the meeting had started, but Jill Dunwich, the Chief of Staff's assistant was now seated between the Vice-President and Fairchild and a young secretary was in a chair before a transcribing machine to the left of the President. The empty lunch plates were mingled between stacks and charts of GAIA information. The President pushed his chair back from the table and turned to the secretary.

"Get your draft of that directly to me when you're done, I'll probably be upstairs when you finish. That'll be all for now." The President turned back to the others.

"Jill, you need to notify the Chief of Staff and set things up for a full cabinet meeting Monday afternoon. Don't tell anyone except your boss what the topic is. We have to maintain tight control of this until we get things in order. And you will see to it that the Commerce Secretary is in my office tomorrow at noon."

"Yes, sir. Assuming the Secretary is in town on a Sunday."

"If he is not in town, then get him back for the Cabinet meeting Monday and I'll talk to him Monday morning."

Jill simply nodded and gathered her papers.

"Bill, you have the National Security staff get to work on a situation report based upon what we have here for the Cabinet meeting. And you'll check with CIA, FEMA and the Pentagon to see if they already have operational analysis or plans for such an environmental crisis already set up."

"Right."

"And you two," indicating the Vice-President and the Senator, "will notify the big five on the Hill of our situation and have them come to the Cabinet meeting. Of course, exhorting them to keep this quiet until we decide how to break it." They nodded.

"By the way, Jill. See to it that our staff gathers all this up," indicating the charts and reports on the table, "and makes enough copies for everyone here and the other staff people who need it to prepare things. Everyone who gets it should know it is classified Secret for now.

"Doctor Fairchild, you coordinate to get everything we need from the people out in Colorado. You might want the top man out there, what's his name, Brady, here for the meeting. Go out there tomorrow if you need to, to get whatever is necessary. You will be briefing the Cabinet first on the situation. You will also coordinate with the Chief's office to set up my trip out there to see the GAIA set-up in the next week or two. Jill you get on that, too." Both Jill and Fairchild indicated they understood.

"Well, that's about it for now. I think we did a good job for such short notice. Let's pray our actions are fruitful." The President smiled and turned to the Oval Office, but turned to Andy before he left, "Andy, can I see you in my office for a moment."

As the President turned, Andy and the Vice-President exchanged glances and they shook hands. Andy picked up his now empty briefcase and followed the President out.

Andy closed the door to the Oval office behind him. The President was already over closing the opposite door. The President walked over to Andy.

"Senator, I guess we both know that you could have used the information on this in a much different way than you did."

Andy nodded and replied, "I think we both know that the way we are proceeding now is the best for our country, too."

The President continued, "I just wanted to acknowledge that my administration's pants could have been down all the way on this and that I really appreciate your bringing it to me. And that you have my gratitude for that."

"Mr. President. I do appreciate you telling me that."

The President and the Senator shook hands.

———

April - Boone, Iowa

Jared Tolliver knew it was still hopeless before he even got across the yard to the machine shed. His rubber boots were already so covered with the thick brown mud that each step was small struggle. If he could not walk to the machine shed without trouble, he certainly could not plant the fields. Looking up, he saw the leaden skies were not promising either.

After the horrible crop results he had last year, he had followed the advice of the state extension agent and moved to a modified no-till method. That had meant a very shallow plowing almost before the ground was thawed, but before the spring rains like those of the last three years came.

The real idea behind "no-till" was to reduce the erosion of topsoil and the extension agents had been trying to sell the farmers on the new ideas for years. But the farmers had resisted; something about the fresh churned soil of a new plowing made the farmers think everything was right for spring planting.

However, this last year they had come up with the idea that the new tillage techniques would also reduce the risk of getting caught without enough time to plow and plant between the time the ground was workable and the pesky spring rains that had become the norm in recent years.

So, Jared had spent the money for a new shallow plow that actually performed more of a scrap and roll of the top inches of soil rather than any kind of plowing he had ever known. You could not plant either corn or soybean too early or the quick sprouting seeds would freeze in the late frosts of Iowa. The corn and soybean farmers of the American Midwest had to wait for the traditional dates for last freezes in their area and hope they had time enough before the rains hit.

Unfortunately, this year had been worse than last year. The last three storm fronts that had swept out of Canada and which usually would have meant March snowstorms, had turned out to be full of thunderstorms and tornadoes. Each storm had dropped one to two inches of rain on the shallow tilled ground and the result was lots of mud and what looked like worse erosion than would have happened if the soil had been broken up to a deeper level, which Jared figured would have held the water better. The expected spring freezes never did come.

Not that his neighbors who had not bitten on "no-till" were in any better shape. They had not even been able to get out to plow yet.

Jared did not even bother opening the machine shed doors. He swung himself up on the cement casement that had served as part of the old dairy barn that had once stood here in his father's days. From the top of the casement he could see all three quarter sections he now farmed.

The southeast quarter seemed better today, but he could see the slight glistening patina to the dirt which his trained eye saw as mud; wet mud. The southwest quarter with its low depression in the middle and left sides still had standing water in it backed up all the way out to the county road on the west. Only the quarter directly to the west showed promise. Perhaps by the weekend it would be dry enough.

Jared Tolliver sighed heavily as he surveyed the morass of his fields. He had 480 acres of what was supposed to be the most productive farmland on earth. Prime Iowa black dirt. And for the third year running it looked like he was going to be delayed in planting it way beyond the point of highest profitability. Another couple of years like last year and he would be done for in farming. Hell, he would be done for this year, if it did not dry out so he could plant something.

His gaze turned from the wet field to the casement he stood on and the remains of the dairy barn piping that still poked out of the cement pad the old barn had stood on. He had torn down the barn and sold the aged wood to some interior decorators from Chicago several years before. He and his sons had not done any dairying since his father had retired in the early eighties and sold off the excess land that they had used for the herd. And he had sold the last beef cattle a few years later when the price of beef had been undercut by the huge cattle feeding operations out west. The traditional multi-faceted Iowa farm had succumbed to stable corn and soybean prices and reliable government programs that encouraged dirt farming on this marvelous soil. Now he lived or died by corn and soybeans.

Perhaps the time had come for him the branch out again. Not to cattle or dairy, that still did not pan out for the small farmer. But, maybe some crop that could flourish in this great soil and would not need to be planted until he could get into the fields after the rains. Or maybe something that could be planted before the last frost. Or, for that matter, maybe something that could go in the autumn before, like winter wheat or hardy sorghum or millet or maybe that newfangled quinoa grain for the New-Agers. As he started back to the house, he felt the first drops of another rain shower on his cheek.

———

April - Washington, D.C.

"Thank you, Doctor Brady, and Doctor Fairchild." the President said as the men finished their briefing of the packed Cabinet Room. He turned his head up and down the table to see how the senior members of his Cabinet had reacted to the briefing. No one spoke.

Arrayed on either side of him at the Cabinet table were the members of his Cabinet, the Secretaries of Defense, State, Treasury, Interior, and the rest. Immediately to the President's right was the Vice-President and the White House Chief of Staff, and to the President's left, the Speaker of the House and the Senate majority and minority leaders.

All around the walls of the room, in matching blue velvet armchairs, were the remaining members of the President's inner circle and others. Andy Knowles sat in one of the chairs, along with the House minority leader and two other members of Congress. Brady and Fairchild sat on either side of a chart easel at the far end of the room. They had left the most thought provoking of the GAIA charts up on the easel as they sat down; the chart showing what the United States would look like with a twenty meter sea-level rise, that which would occur if all of the polar icecaps melted. They had made it clear that this chart was not a likely occurrence in their lifetime and a remote possibility beyond that, but seeing the chart with huge chunks of American seaboard missing for the first time certainly made you stop and think, Andy remembered it only too well.

"Mr. President?" Admiral Kirksey raised his hand from the opposite end of the table.

"Yes, Admiral."

"I wanted to add one important thing to what these gentlemen have presented." Kirksey spoke slowly. "We have received word that Russia has issued orders calling for the consolidation of emergency communications and disaster response personnel in certain areas. At first it seemed to be in response to the local flooding common in Russia every spring, but with the information we have just heard, plus certain other reliable information we believe Russians are reacting to coastal flooding. It seems that three of the Russians' primary port areas, the Gulf of Finland, the Sea of Azov on the Black Sea and the military ports on the White Sea are all on geographically enclosed areas, which, because of tidal forces on elongated tidal basins have had the effect of the sea-level increases we have heard of being enhanced, this tidal effect is only complicated by the normal spring melt of snow and ice. As of this morning, the U.S. Consulate in St. Petersburg reports the section of the city near the port area and the Neva River are under from two to five feet of water."

"It seems we are not alone in our concern for this problem. Thank you for the information." The President shifted in his seat and began speaking.

"I realize many of you will have a lot of questions. Dr. Brady will be available for further questions when we finish with the meeting and the White House and N.O.A.A. will have reports out to everyone here today plus some others in a few days. I think you can see we have a situation of cataclysmic impact facing us. I see our duty at this time to be threefold." the President paused for a moment.

"First, we need to put everything we have into finding out exactly what is happening, and how fast. Second, obviously, is to find out what we can do to mitigate the damage that has already been done to the world environment and find out what needs to be done to stabilize things. Lastly, we need to make sure that what we do in reaction to this is not itself damaging, to both our people and the people of the world."

The president continued, "If we do not handle this right, the news itself could be devastating, it may be that no matter what we do. It could cause widespread panic, financial collapse and God knows what other mischief, both in our country and abroad.

"It is now three o'clock eastern time. At four-thirty, I want Treasury, S.E.C. and whoever else is involved to declare, on presidential orders, a two day banking holiday. All securities and banking transactions will be halted until Thursday morning. That should give us time to catch wind of any big disasters coming our way in the markets. If that isn't enough time, get back to me, but I don't think we can expect John Q. Public to take any more than two days without banking. Treasury?" the President looked across to the Secretary of the Treasury. "I want you to coordinate the financial aspects along with the Chairman of the Fed and the Council of Economic Advisors. When we start talking about the coastal problem, we may have some major banking and insurance companies hanging out a long way in both coverage and investment. Not to mention the cities themselves. And the agricultural impact could have drastic repercussions on the commodities markets. Put your best and brightest on it, and get me a report by Wednesday at three as to what the effect of the announcement I am going to make will have had and what we can do to cushion the impact." The Secretary nodded.

"I guess I got ahead of myself there, as to my first two priorities," the President paused and looked at his notes. "I see the need of finding out what is happening and finding out what to do about it as being two halves of the same problem. On the policy making level, I am going to utilize the existing Council on Environmental Quality as my advisory and policy planning staff. But Doctor Fairchild is going to revamp it so that I have the best people in the country on the Council to advise me and this Cabinet on the environmental concerns facing us. As my first appointment, Dr. Brady," the President peered over heads toward where Ken Brady sat, "you will help Doctor Fairchild, get the right people on the Council." Brady nodded and looked over to Fairchild.

Again the President checked his notes and continued. "I see the traditional roles of the Environmental Protection Agency and N.O.A.A. as being integral to our dealing with this situation. And, in this emergency, I can't see having to go to one agency to find out about the cause and the other to find out about the effect. Insomuch as the position of Director of N.O.A.A. is vacant, as of this morning," the President's glare around the room stifled the murmur at this announcement, "I am going to ask our E.P.A. chief, Ruth Anne Warren, to take over executive duties for N.O.A.A. until Congress can put through legislation to make N.O.A.A. a part of E.P.A." The President looked over to Mrs. Warren, who swallowed hard and nodded.

"Senator Knowles?" the President looked at Andy. "Since that idea was yours, I trust that you will get it through Congress for me, and quickly."

Andy smiled and nodded. He wondered how many of his other suggestions were going to be implemented.

"Finishing up on that piece of business. The Secretary of Commerce has also decided to, er, retire, effective immediately and I will nominate a replacement as soon as possible." The President let the murmurs die out as he sorted his papers again.

"I will go on national television at nine tonight and outline the situation for the people. I won't use any of the worst case scenarios we have seen. No use in scaring the crap out of everyone. But I will use the best information we have on most likely outcome and let the people know we have a serious problem.

"I will also outline a skeleton of a national policy to combat the threat facing us. Certain products, including the worst of the greenhouse gases will be banned as soon as possible. New coal fired and other fossil fuel industrial facilities without carbon-dioxide scrubbers will be banned. I understand that all the major manufacturers have developed a range of electric and hybrid vehicles, but various economic and market conditions haven't made a significant shift in sales; well, we will make such efficient cars immediately feasible by regulatory action and tax incentives and if necessary, import penalties or other pressures on the auto industry. Also, the nation's timber cutting programs will be reviewed and sharply reduced." The President glanced quickly to the Senate minority leader, who was from the Pacific Northwest. He turned to the Treasury Secretary. "Those are bound to have market impacts. Hmmmh?"

"Continuing... I will announce that henceforth, all U.S. foreign aid and our vote in the World Bank, the U.N. and elsewhere will be tied to compliance with environmental concerns. The developing countries won't be getting billions in aid from us to help them cut down the rainforests or build dirty factories and choke off our oxygen. That's why we have to cut the hypocrisy and rein in our own timber cutting and other industries. The same foreign aid policy will apply to industrialization and chemicals. Further, it will be the policy of the United States to take proactive steps, including economic embargo and, if necessary, military intervention, to prevent actions by foreign powers which directly damage the world environment. The State Department will forward a synopsis of my speech to all foreign governments."

The President turned to the Energy Secretary. "I am going to announce that the Energy Department is undertaking a FIVE year plan to shift our nation's energy reliance from coal-oil-hydrocarbon based energy to solar, hydroelectric, tidal and other renewable energy sources, including research into whether nuclear energy can be done safely. We cannot wait for the ten and twenty year time frames we previously had as goals. This will coincide with a moratorium on all new leases for oil exploration and development on federal land and a prohibition of permits for any energy project, or any industrial project, which has adverse impacts on the global or national environment."

"On the domestic front, I will propose and expansion of our national service program, which will provide man-power, that is man and woman power, to mitigate the damage to the environment. The Departments of Agriculture, Interior, Housing and Urban Affairs and others will all have to have programs in place to provide housing, resettle people, plant forests, feed people, etc., and this corps of young people, the Vice-president called it the Earth Corps when she and Senator Knowles suggested it to me, will be the ones who carry this out."

"I know I've hit you with a lot this afternoon. Believe me that I understand your reaction. I felt the same, until I studied the information before us and saw that it is valid and needs immediate action. We have hidden our heads in the sand for decades now on these environment concerns and we are going to be called to task for our ignorance and arrogance.

"I want weekly reports from every department and agency in the government on the status, predictions, policies and ideas for this program we are undertaking. Direct your reports to the Chief of Staff with copies to Dr. Fairchild and the Council on Environmental Quality and whoever else you think needs the report. We will have another full Cabinet meeting late on Wednesday afternoon, after we see what the reaction to all of this is. I will be calling individual Cabinet members in to talk over the next few days, please make a point on staying handy, in case I need to talk with you."

The President had now stacked his papers in a neat pile before him, he opened his mouth to speak and then waited, as if in thought, "Gentlemen, and ladies, I hope you do not find it melodramatic for me to say that I feel we are at one of the great watersheds of history. What we do at this point in history can affect all future generations in the most profound way. Of course, it is not just our country that is in this position, but we in the United States are in the unique position of having the knowledge, the technology, the resources and the influence with other countries to do something about the disastrous events facing us. Hopefully, we also must have the will and sense of purpose to make the hard decisions that will have to be made.

"It is precisely because the stakes are so high for both us and future generations that I must ask everyone from my Cabinet, to the leaders of Congress assembled with us here, to put away political differences and, most certainly, regional differences to act as a nation in a concerted effort to set our planet right again and protect life as we know it on Earth."

April - Miami, Florida

Nobody really knew how it got started. Some blamed the Miami Herald for bannering a headline of "MIAMI UNDER WATER?" in the late edition in response to the President's speech. Others blamed an over reaction by the Miami Police and Dade County Sheriff to a sudden influx of people, many Cuban and Haitian, to the downtown bus station and the airport. Still others pointed to the President's unprecedented banking holiday, which had caught everyone unawares and resulted in many automated teller machines, financial institutions and other businesses being closed or out off money. Many community leaders saw the rioting as an expected reaction to the already unstable racial, ethnic and economic hodgepodge of Miami, Florida meeting with a flurry of bad news and inopportune events.

The governor had responded with a call-up of national guard in the early morning hours to combat the looting and violence, which was particularly bad between the poorest of the Cuban and black neighborhoods. With the National Guard trucks and jeeps moving through residential neighborhoods, word had quickly spread through the masses of partially literate and misinformed people who only knew the rudiments of the message the President had tried to convey on TV the night before, that the Army was evacuating Miami.

Before the media and civic leaders could sift through the events and try to pass on correct information, thousands of families had fled in the night, clogging the freeways north. The empty houses only exacerbated the looting problem.

By dawn's light the morning after the President's speech Miami was in chaos. Those who had fled, but were returning and those who still thought they must flee, either from the misunderstood rising ocean or from the very real danger of violence, had the streets and highways choked off. On three sides of the city massive riots were taking place. Some pitted minorities and against police and Guardsmen. Some riots had Cuban gangs taking on Haitian mobs in pitched battles.

The worst of the fires started in Little Havana. Molotov cocktails aimed at absentee business owners or simply thrown in frustration or fear started fires which the firemen refused to go in and put out.

By nightfall, all those who cared to listen had been informed by community leaders, the governor and even the President that there was no immediate need to fear that Miami was going to be flooded and what the President had spoken of was theoretical, if the world did no t do something about the environmental problems. But for many Miami residents, the fabric of civilization had come unstrung. Neighborhood strife continued unabated.

A well-placed fire bomb in an electrical transformer yard knocked out electricity for all of central and south-central Miami just before sunset. Cable television for Greater Miami went out with the central electrical network. The purveyors of cable television would be given an object lesson on how pervasive their service had become when the utility workers could not get electricity back on and all of Miami was without their usual news source, television. Unfortunately, the television had lasted just long enough to get the word out about the looting and violence.

As darkness, violence, misinformation and fear spread throughout the city, the violence and destruction of the previous night escalated. Police and Guardsmen struggled to confront the worst of the mass violence, unable to respond to individual incidents.

Youth gangs and drug dealers, already armed to the teeth for turf battles and drug running, took advantage of the situation to conduct brazen, organized raids on unfortunate merchants, electronics stores and other targets of opportunity in the blacked out areas. Old vendettas were carried out with remarkable ease in a city without light or civic organization and in which harried police forces struggled to maintain their own safety on the streets.

———

April - Washington, D.C.

Andy Knowles waited while his staff distributed copies of the press release and media kit to the press corps. Bill Jolson, and Dee Dee Lindquist were circulating with the handout. Several Senators and Congressman stood on either side of Andy. When the staff members were done, Andy had to wait for one television technician to adjust a microphone on the podium and get out of the way. The Senate Media Room was packed.

"Hh-Hmm." Andy cleared his throat, paused just long enough to insure that the throat clearing would be edited out of any TV clip, then began, "We have handed out several items concerning emergency legislation which will be introduced to deal with the environmental holocaust we are now facing.

"Included in these provisions are several things which the President mentioned in his address last Monday evening. I agree with President's actions, and I believe most of my fellow members of Congress, in large part, do also. At times like these we see no reason for partisan politics to interfere with taking the all-important actions that are needed to set our world environment right.

"Additionally, I have proposed a twenty-five percent tax on the importation and production of fossil fuels. The proceeds of this tax would pay for the many public service projects, such as the sea-walls and levees to protect coastal cities from flooding and the Earth Corps and other ecologically important programs, including a program to implement CO_2 scrubbers for all appropriate industry.

"Also, I am proposing an immediate cessation in the production of dangerous fluorocarbons. For some time we have had plans for phasing out these dangerous products. We can no longer wait for a phase out. We have many suitable alternatives for most of the industrial and refrigeration applications of fluorocarbons. I believe that it is a simple matter of economics as to why American industry has not already ceased this dangerous, no, deadly, poisoning of the earth's upper air with atmospherically reactive fluorocarbons.

"Included in the package the President requested which will merge the functions of the Environmental Protection Agency and the National Atmospheric and Oceanographic Administration, I have included a supplemental appropriation for certain federal agencies, notably the Department of Defense, Interior and the Coast Guard, to undertake the immediate repairs, protective actions and research to protect our coastlines from the high sea levels we have been experiencing. Even though we may be able to take immediate action, both in our country and around the world, to change our habits and reverse the pollution and global warming, many of our coastal areas may be irreparably harmed by high water right now. Our coastal estuaries, wetlands and beaches, not to mention our private property along the coasts need to be protected and this legislation will give these agencies immediate resources to start this task.

"There are also included in the package, appropriations for the Housing and Urban Development and the Agriculture Departments which will allow these agencies to study and make plans for any environmental changes which may affect the way America lives and produces its food.

"The unfortunate civil disturbances in Florida are certain proof that our citizens are concerned and frightened by what is happening to our planet. We, in Congress, need to take immediate steps, not only to fix the environment before it is too late, but also to protect our citizens from the ravages which our past transgressions have already caused."

Andy now stood aside. One by one the other Senators and Representatives on Andy's flanks spoke to the press, but it was clear to all present that the show was Andrew Knowles', all the way. He had had nearly two weeks to prepare the emergency legislation based upon both the GAIA team's materials and his contact with the President, but also on the many previously proposed environmental measures which had fallen victim to fiscal restraint and a general disbelief in the problem.

After all were finished, the press was given an opportunity for questions. The Cable News reporter went first with a question to Andy.

"Senator Knowles, in regards to your proposal to tax oil twenty five percent, don't you think that is a little radical? Do you really think you have a chance of getting something like that passed?"

"Virtually every other country in the world has significant taxation of both imported and domestic energy production. For years we, in America, have been taking a free ride in our gas-guzzling cars and petroleum based industry at the expense of our environment. If we are going to be able to win this fight to save the environment we will have to ask other countries to make hard decisions. Our country needs to make this decision to bring itself in line with reality and to keep ourselves from appearing as total hypocrites to the emerging nations of the world who we are going to have to ask to stop deforestation and industrialization until the ecosystem's problems can be dealt with. If the rest of the world were as self-centered, pollution-ridden and wasteful of energy as America, our world would be on the verge of dying right now. Assuming it isn't. That's still to be seen.

"If we don't start taxing our wasteful energy habits and start doing something about our reliance on non-renewable energy, our children may not be able to import oil because the oil tankering facilities will be under water, as will half of our major cities, much of our industry, most of our nuclear power facilities."

"But, you can't...." the cable reporter cut in.

"I'm not finished." Andy was staring into the cable camera. "If you ask me whether I will let a question of what is politically expedient and financially comfortable to Wall Street affect my judgment as to whether we should continue on a course of action which might starve millions of Americans. I think you know my answer. If you ask me whether I will let this year's oil company profits dictate whether Houston is under the waters of the Gulf of Mexico by the year 2050. Again I think you know the answer. If you ask me whether I will let the price of gasoline and the increased cost of making energy efficient automobiles decide whether the very balance of life on earth is upset. Then, I certainly hope you know my answer.

"The problem is that we, the American people, the people of the world, have been ignoring these questions for too long. What the President was trying to say is that our best scientists are telling us that the time to answer these questions is now. If it is not too late. Either we start making the hard decisions to not only tax petroleum, but also to live within our means, be frugal with resources, cherish our planet; or the decisions will be taken out of our hands and we, our children and our posterity will long regret it."

There was a pause after Andy spoke and then the Congresswoman from Colorado started clapping. She was joined by others and soon the media corps. Andy looked around, mildly surprised. Ovations at Capital Hill press conferences were indeed unusual. Andy savored the occasion and then decided to take advantage of the moment. He folded his notes and left.

———

April - Miami, Florida

"Every cloud has a silver lining. Eh, Raul?"

"No shit. Billy. That was one bitching idea you had."

The two men sat at the kitchen table. Billy was stacking the cash in piles. Raul was using the apothecary scale to measure out the white powder and then using the freezer bag sealer to neatly finish off the top of the small bag that would sell for one hundred dollars.

"And the best part is that we don't have to tell Carlos anything about it. Do we?" Billy asked.

Raul looked up at Billy, his face ashen, "You sure that's smart? Carlos don't take no shit."

"Of course it is right. It was my idea to knock over those fucking Cubańos. I figured it out. I set it up. And you and I pulled it off. It just wasn't part of Carlos' action."

"I guess."

The two men worked in silence for a minute, until Raul let out a burst of laughter and said, "We sure did kick ass though, didn't we? How the hell did you know the Cubańos were having a deal go down?"

"Shit, that was just luck. I was going for the drugs that I knew they had just picked up from my fellow Colombians. It was just pure luck that they had a buy going down when we hit them. I guess they thought the buy would be as safe and easy with the power out as we did our hit."

"Did you see the look on their faces when we popped them."

"Yeh, almost as good as the look on your face when you opened the second briefcase and saw the cash."

Both men laughed. Billy tapped the last stack of cash on the table to square up the edges and set it on the table, joining the dozens of other stacks.

Raul looked at him and asked, "How much is there?"

"Each pile is a hundred 100's," Guillermo "Billy" Barraca said with a smile.

Raul looked at the ceiling as he tried to multiply one hundred times one hundred. Finished, he looked back at the other end of the table, dumbfounded at their good fortune. Their early morning raid into darkened, riot-torn Little Havana had netted them several hundred thousand in cash and slightly more, street value, in cocaine.

By noon on the third day the police, state officials and military were able to restore a semblance of control to Miami. An uneasy calm settled over the city. Damage from fire, theft and riot was estimated in the hundreds of millions of dollars. Deaths were in the hundreds and lopsidedly heavy in the minorities. Like the Los Angeles riots after the Rodney King verdict, the Miami cataclysm had been set off by an event that soon became totally unrelated to its effects.

Reports and rumors of surrounded and outnumbered police units opening fire on groups of civilians were everywhere, as were reports of intentional attacks on police by armed gangs. Some were true and some were not. Both truth and rumor fueled bitterness and hatred on both sides.

As Miami strained to heal its wounds, it was clear that some wounds were permanent. Some of the social structure of the city had been torn asunder. As the city tried to rebuild, there was always an ominous overtone to every action. The President's warning about the GAIA team's hypothesis overshadowed everything. The fact remained that the city of Miami was built on reclaimed swamp and tidelands, all of it only a few feet above sea level; some housing developments were actually below sea-level and relied on dikes that kept the sea at bay. It became readily apparent that out-of-state lenders and insurers were not eager to put their money on the line in Miami. The same was also true for some of the low-lying cities not hit by riot like Miami. New Orleans, Houston and several others cities found that funding sources were disappearing and municipal credit ratings were slipping radically in the wake of the Miami events. The states of Florida and Louisiana, with their biggest cities all sitting in the shaded area of the GAIA teams inundation prediction maps found bond ratings slip to the point that traditional government financing was impossible.

Above all of the other results of the Miami experience, the government, from the President on down, learned one overpowering lesson. Be careful when you warn the populace of impending environmental catastrophe. The President's attempt to insulate the country from the dangers of the environmental bad news had clearly backfired. It was the unfortunate legacy of the Miami riots that politicians at all levels became leery of giving public notice of environmental warnings.

———

May - Dhaka, Bangladesh

Much of the prime farmland of the nation of Bangladesh had been inundated with the saltwater tidal surge of the typhoon. And many fields became no more than briny pools, for, in some areas, the waters did not recede to pre-storm levels. This was attributed to the flooding and sedimentation of the natural drainage channels. Fully a third of the nation is less than five feet above sea level and this area is home to forty million people. In spite of the best efforts of the world community, millions would die from starvation from failed crops and waterborne disease.

Trying to count the dead in Bangladesh became a useless waste of time. The pre-disaster census of the overcrowded nation had been a guess on the part of the United Nations and placed the nation as the world's eighth most populous. To try to count the millions who died outright from the storm or the millions who died in its aftermath became an impossibility. One estimate projected that as many died in the Bangladeshi typhoon as had died world-wide in World War II. No one had reason to argue with the estimate. It was clearly the most deadly natural, or man-made, disaster the modern world had ever known.

Akim Darpheel stayed at his post as long as he could. However, it was obvious that there was no purpose in remaining in the general staff building. The three main garrisons of the Bangladeshi army had been utterly destroyed in the storm. The brigadier in charge of the Chittagong Barracks had mutinied and declared independence and was immediately supported by Indian Army units out of Manipur, seeking to coalesce control over the only route to the sea from the Indian territories of Manipur and Tripura. The Indians were apparently taking advantage of the destruction to recover parts of Bengal severed from it nearly fifty years before.

The General Staff was a headquarters without an army. Like his brethren, both officer and enlisted, Darpheel had thoughts of trying to escape from the general staff building and perhaps Dhaka.

They had tried to work to coordinate relief and rescue, but the sheer enormity of the devastation had overwhelmed the Bangladesh government. Most medical supplies ran out in the second week and food supplies dwindled shortly thereafter. The attempts by other nations to assist were stymied by a lack of transportation, both into and across the ravaged country.

In Dhaka, the starving, homeless masses assembled in the central square where the Army had tried to set up the aid stations Darpheel had seen on his trip back into headquarters. But too many people converged on the aid centers and overtaxed the ability to resupply. Cholera became rampant in the still submerged areas of the city and in the submerged areas of the countryside those who did not have the disease were the exception. Foreign aircraft refused to land at the Dhaka airport due to the disease and likelihood of danger from the mobs of homeless, starving people. Pure water was rapidly running out in the general staff building.

As hunger and death mounted, so did anger, rioting and violence. The parliament building and several other buildings were ransacked and burned. The soldiers manning the general staff building held out as long as they could, but their supplies were shrinking rapidly.

The generals and their families had flown out by helicopter at noon. The landing of the helicopters had created another of the riots that had become so common. The masses in the government plaza had thought the landings were new food shipments and had stormed the general staff building when the aircraft landed in the walled enclosure. Like earlier riots, only tear gas, concussion grenades and, finally, machine gun fire had turned the masses back. Piles of bodies blocked the entrances and the plaza in front of the building.

No soldiers left the building to remove the bodies or tend the injured. The bulldozers used to bury the dead had run out of fuel. There was no one and nothing to tend the injured with.

The armorers were issuing sidearms to all officers and Chinese assault rifles to enlisted. Akim took an automatic pistol and two clips. As he left the armory, he noticed his superior in the Training Directorate, Major Radija, directing a work crew carrying crates to the rear loading dock. Some of the workers were obviously soldiers, but were in civilian dress. He followed Radija to the docks.

Two large trucks, the big American two and a half ton vehicles, which were full of soldiers and what appeared to be their families were backed to the loading docks. As he watched, Radija and Colonel Majlis directed the loading of the last crates. The workers also jumped on the trucks and clanged the gates shut.

Major Radija, seeing Darpheel motioned him to the trucks.

"Well, are you coming or not?" Radija asked him.

"Where to?" Akim asked his boss, incredulous at what was taking place.

"Out! Away! Anywhere. Things are hopeless here. The city is in chaos. There is nothing we can do. We'll head north. You're from the Assam border. It is better there. Hurry!"

Radija motioned Captain Darpheel onto the truck as he went to the cab. The colonel helped his wife into the other truck's cab and climbed in after her.

As Darpheel jumped into the Major's truck, the truck with the colonel rolled toward the back gates of the general staff complex. Darpheel took a seat on a wooden crate in the middle of the truck.

Soldiers and their families crowded the truck, perhaps twenty-five in all. Among them was the little half-breed sergeant, Quist, who had gone with Akim to the delta. As the truck cleared the back gate, the man who had opened the gate ran over and swung himself over the tailgate and aboard the truck. He muttered a word of praise to Allah as he flopped onto the floor at Darpheel's feet. The back gate to Army general staff headquarters was now unguarded. The Bangladeshi Army was abandoning the capital. The two non-commissioned officers sitting at the tail gate rammed the bolts closed on their assault rifles.

They encountered no trouble during the first few blocks out of the gate. Darpheel noticed the putrid stench of death he had come to recognize, both in the delta and in the last few weeks in Dhaka.

As they reached the main boulevard they could hear the shouts and murmur of a crowd. The trucks turned north up the boulevard and picked up speed. Through the rear opening in the canvas cover of the truck Darpheel could see people running in the trucks' wake.

The drivers were plowing through the crowd to prevent the trucks from being overtaken by the thousands of people along the road. An occasional bump, followed by a body back in their tracks showed that the colonel and the major were sparing nothing to escape the city. An occasional runner managed to grab the tailgate, only to be met with the butt of the non-coms' rifles.

Captain Darpheel closed his eyes and mumbled a prayer. He had gotten much more religious in recent weeks.

———

May - Washington, D.C.

"This morning's session is adjourned. The Senate Committee on the Environment and Public Works will reconvene at one-thirty this afternoon, when our witnesses will be representatives of the farming and manufacturing industries." As he finished speaking, Andy cracked the gavel down.

As he gathered up his papers from the morning session, Andy saw Derek winding his way through the room to him.

"There is a change in schedule for lunch. Hudspeth called and insisted on meeting with you before he would clear the funding bills out of his committee. I thought it was important so I got Rich Claypool to cover your luncheon speech. He needs all the national coverage he can get for the re-election next fall, so he didn't mind. In fact, he thanked me for thinking of him, like we were doing him a favor." Both Derek Shaw and Andy Knowles smiled at the eagerness of the first term Congressman who now held the seat in Connecticut Andy had first been elected to. Derek handed a paper to Andy.

Andy stifled a yawn and looked at the paper. Derek continued speaking as they walked out of the committee room.

"This is a list Dee Dee made of the funding bills Hudspeth has control of in his committee. Virtually all of the environmental cleanup funding, the Earth Corps money, the seawall fund and the foreign aid funding measures for overseas programs, just to mention a few.

"Of course, he is concerned about the forestry and fishing restrictions that he really doesn't have any control over, since they aren't coming through his committee. Grapevine has it that both Hudspeth and the Minority Leader are putting everything they have into stopping the forestry and fisheries..."

Andy cut Derek off with a wave of his hand and said, "Right, right. I get the message. Hudspeth is going to have an old fashioned arm-twisting session with me at lunch. So what bargaining chips do we have?"

Derek delayed answering Andy while they passed a group of journalists headed the other way through the tunnel back to the Hart Senate Office building. He continued.

"Of course, the fishing bill is the least important as far as the environmental package goes. It really isn't essential at all. But State asked for it to use as leverage on the Japanese and Koreans on other more important environmental concerns. It does have some ocean pollution and oil spill measures that could be significant improvements, but those really aren't what Hudspeth and company are worried about. You could agree to split the fisheries bill up or delay the parts which could effect jobs in the domestic fisheries industry. The big stumbling block is the forestry restrictions. Hudspeth doesn't want to appear to give in to the folks back home and getting our own forestry usage in control is essential to putting pressure on Brazil, Indonesia and the other rain forest nations to control the deforestation."

They had reached the fork in the tunnel were Andy would head to the right to the Capitol for lunch in the dining room and Derek had to turn into the Russell Building. Andy handed the committee papers to Derek and said, "Well, wish me luck and keep the liniment handy in case the arm twisting gets too heavy." He turned towards the Capitol and walked off.

Senator Harold Hudspeth did not look like he was a man who controlled the purse strings of the nation. An elderly, plump man in a rumpled suit whose sparse hair always was in disarray, Hudspeth appeared to Andy to be more of the small town lawyer sort or perhaps a union boss. But then, Andy's perception of the powerful Senator was jaundiced by the knowledge that a small town lawyer was exactly what Hudspeth had been before he had put in place a political balance between Oregon liberals and Portland trade unions which had elected him to national office before Andy had gone to college and had returned Hudspeth to Washington every election since.

The seniority system kept Harold Hudspeth in place in his powerful Budget Committee chairmanship and his chairmanship kept him in control of virtually every Senate bill involving money. His strangle hold on the purse strings and concurrent ability to get things done for his constituents, benefactors and whomever else played things his way was what kept him in office. And it was what made it necessary for people like Andrew Knowles to have lunch with the old man from time to time.

Andy crossed the dining room, nodding to a few people as he passed. Hudspeth was seated on the far side. He was already munching on his salad. Hudspeth looked up as Andy approached.

"Ah, Andy, my boy, glad you could make it." Hudspeth rose and shook Andy's hand.

Andy cringed at being called "Andy, my boy" by a fellow Senator, but realized that this was part of the old senator's persona. Andy returned the hearty handshake.

"Good afternoon, Senator." Andy said as he extracted his hand from Hudspeth's grasp.

The waiter came over and took Andy's order, Hudspeth had already ordered. After a few polite preliminaries the old Senator got right down to business.

"You know, Andy, we have quite a few of these environmental measures, both from you and the other stuff the President proposed at the Committee and I thought it would be wise for us to have a chat, just so we are on the same sheet of music."

"I think that is a good idea. I must say that I am a bit concerned that it has been six or seven weeks since the GAIA information was released and we still don't have any significant legislative solutions out of committee yet. I wouldn't want anyone to be able to make any political hay by pinning the blame for anything that happens on our party. The people are really concerned about this and the few changes the President has been able to make by executive order are all that they see happening. I wouldn't want to lose the ball on the environment. It's just too big."

Senator Hudspeth put his salad fork down and leaned toward Andy. "Well, you're right. The people are concerned. But they are also concerned about jobs. I want to protect the environment as much as you, Andy. But, I want to make sure that we don't walk over the little people on the way to do it."

Andy thought, 'Yeh, little people like Carolina-Pacific and Northwest Paper Products Corporation!' He, instead, said, "Exactly what were your concerns, Senator?"

Senator Hudspeth started to speak, but he was interrupted by the arrival of the lunch plates. After a few bites, the old senator turned back to business.

"Let's not beat around the bush, Senator. We both know what we are here to talk about. The timber industry in the Pacific Northwest simply cannot survive under the conditions imposed by your environmental package, at least not just yet. And the effect on the fisheries is almost as bad. If you want my support for your environmental program you're going to have to give some leeway in these areas."

Derek had been right on the mark, Andy thought. Andy thought for a moment and then spoke, "I think we can work on the fisheries regulation, that isn't a critical part of the package. But the problem of deforestation is a big concern for everyone. It appears to be a major reason for the carbon dioxide shift that is behind the global warming. We can't very well force the foreign countries who are hell-bent on cutting the rain forests to stop unless we get our own country in order."

"So you want to destroy the economic livelihood of more than half the counties in my state, just for window dressing to stop the rain forest clearance that is the real culprit?"

"It's more than window dressing and you know it. You know the kind of clear-cutting that is going on." Andy stopped, arguing would get them nowhere, or worse. Before Hudspeth could speak, Andy asked, "Do you have any suggestions for what we can do about the forestry controls?"

Senator Harold Hudspeth smiled broadly and said, "I'm glad you asked that." He pulled a marked up copy of Knowles' forestry bill from his breast pocket.

It was now Andy's turn to smile. He hoped the smile did not reflect the sarcasm and irony he felt. Here was one of the most powerful men in the country, holding the earth's environmental health hostage in return for the economic well-being of one industry in his home state. That was exactly what everyone had been doing for years. Andrew Knowles wondered what it would take to wake people up. He was amazed that the GAIA information had not done that.

So, over lunch, the two senators would negotiate a solution. The older senator would get a delay in the restrictions on the forestry industry. The younger senator would get the funding for his environmental program freed up. Hopefully nature would wait for the politicians to finish negotiations.

———

May - near the Garo Hills, Bangladesh

The Darpheel farm was in the highlands east of where the Kangsa River comes down out of the Garo Hills. It was not more than ten miles from the border of India's Meghalaya State.

Mahmal Darpheel's ancestors had farmed the land since before the British had colonized this part of the region known as Assam in the late 1700's. In 1937, Mahmal's grandfather had been the recipient of a bequest from the last of the Muslim rajis and had become the legal owner of the ancestral farms. Growing tea and a variety of spices and fruits, the fertile hillsides had made the family wealthy, at least relative to much of the rest of the nation.

In the safety of the remote hills, the Darpheel family had weathered the nationalistic fervor for independence from the British Empire, the ethnic strife that had followed the British decision to give India its independence and the divisive forces which split Pakistan and India and later East Pakistan from West and had set up the nation of Bangladesh. In their safe haven, the Darpheel family had flourished, giving Mahmal and his brother many sons, and daughters too.

Now his youngest son had returned to the farms. Akim had left them some ten years before, to go the Dhaka and the polytechnic college. As the youngest son Akim would have had no place on the farm, only the two oldest sons would inherit the land, as was tradition. Mahmal and his brother had done so. It was the only way to maintain the farm and keep it workable. Even so, the size of farms diminished as generations passed.

Akim had gone to the college and had become an Army officer, both sources of pride for his father and family, both were unique amongst the farmers in the Assam hills. But now Akim returned to the farm in a tattered uniform and gaunt from hunger. And he was accompanied by several equally gaunt and hungry looking fellow refugees from Dhaka. It was not what Mahmal had hoped for.

The two truck convoy Akim Darpheel had joined had been stopped at a bridge north of Dhaka. The roads were choked with people eager to flee the growing desperation in the city. Some took vehicles piled with their worldly goods, others simply carried whatever they could on their backs. Many wandered with only the scant clothes they wore. Once they left the main part of the city, they did not have to contend with riotous crowds which tried to jump the trucks, but everywhere there were people begging for rides or food. The first major problem occurred at the bridge.

A provincial police official had apparently tried to keep the hordes from Dhaka from pouring into his province. He had blocked the bridge and had a huge crowd gathered on the near side of one of the Brahmaputra tributaries which wound through this part of the countryside.

As the trucks approached the bridge the masses of people parted to let them through. Behind the vehicles the crowd merged back into the road. The trucks pulled to a stop. Through the back of the truck, Akim could see the forlorn faces of the refugees only a few paces away. Hundreds of them, perhaps thousands. The two sergeants stood and brandished their weapons. Akim felt his hip for the pistol.

Akim and his fellow passengers could hear the doors of the truck cabs open and close. Soon they could hear the colonel shouting at someone. They heard only bits and pieces of what was said, "...military business..." and "You have no authority..."

The expressions on the faces of the crowd they could see behind them changed from curiosity to fear. Those on the ground who had room to move shuffled back from the trucks. They heard more shouting and then, the slamming home of rifle bolts.

The voice of the colonel boomed out loud, "I order you to stand aside! This military caravan will not submit to search or other interference by civilian authority."

A long moment of silence was finally broken by a single report from a pistol. Then two more. The crowd cowered back and fell to the ground, taking cover as they could. Finally automatic weapons fire crackled on both sides of the trucks.

Everyone on board the trucks ducked to the floor. The two sergeants at the tailgate peered over the rim.

More shouts and the truck engines revved and gears gnashed. Truck doors slammed and the trucks lurched forward. There was a crash of wood breaking and then a grinding creak of metal on metal from an impact of the truck against what sounded like a vehicle.

Through the back door flaps, Akim could see the bodies of several men lying near the road. All except for one had the gray uniforms of policemen. One appeared to be in an Army officer's uniform. He saw a blue police vehicle flipped on its side where the truck had hit it and a striped drawgate with its arm smashed to the side. Akim could barely see in the dust of the trucks' wake.

They were now rolling over the bumpy plank surface of the bridge. As they watched two more figures in gray ran out of the guard shack beside the end of the bridge. One leveled a pistol at the departing trucks.

"Down! Everyone down!" Akim shouted as he flopped onto the floor.

The pop of the pistol was followed instantly by a gasping wheeze in the truck. Akim felt a body drop beside him on the floor of the truck. He rolled over to be confronted with the face of a private who had been sitting on one of the jumpseats.

The private's eyes blinked quickly. He tried to speak but no sound emerged, except a gurgling. The gurgling was followed by a spurt of blood from the man's nose and then bubbles of blood from his lips.

The bump of the bridge planks stopped and Akim reasoned they were far enough away to be out of pistol range. He sat up and looked back to the bridge. The policeman with the gun could not be seen, the crowds of people behind him had taken advantage of the occasion to rush the guard post and cross the river. Akim turned to roll the private on his back. He knew from the limp body that the man was dead. Akim took the uniform hat off of the dead man and placed it over the man's face.

Everyone in the truck looked from the dead man to Akim. No one spoke.

The trucks continued down the road at high speed for perhaps another mile, then they pulled to the side of the road. The road here was clear of refugees. The police blockade, as it turned out to be, had done its job in keeping this area free from the hordes that had crowded the roads nearer Dhaka.

At the side of the road they took the dead body off of the truck. They learned from the drivers and officers in the truck cabs that there had been an officer shot at the bridge. Major Radija was dead. The major's widow, hugging two small children, sobbed inside the truck cab, the colonel's wife trying to console her. The colonel ordered two men to dig a shallow grave for the private and to be quick about it, they had to be on their way.

After that short stop they drove for several more hours, stopping only once to pour more of the petrol cans into the trucks. At the railroad bridge at Mymensingh they had stopped to rest for the night and make plans.

They had parked the trucks in a coal yard near the railroad siding which had fences and which the colonel declared to be defendable from the hoards of refugees who might catch them in nighttime, and perhaps the local police who may be looking for an Army patrol in two trucks. The colonel assigned men to guard in watches throughout the night. The women prepared the food they had carried with them from Dhaka. They had loaded the first food they had found in the storerooms of the general staff building. Now it seemed ironic that they had picked up the luxury stores stocked for the generals. They feasted on canned beef, sweets and other delicacies that few except perhaps the colonel were used to.

Akim leaned back sipping a bottled lemonade, feeling satiated for the first time in weeks. Everyone else was gathered around the fire discussing plans and debating the wisdom of their next move.

Unlike the rest of his party who debated going on up the Brahmaputra valley or turning east toward the fertile lowlands,

Akim knew exactly where he would go. His family's farm was only a short distance up into the hills. Compared to the devastation of the delta and the living hell Dhaka had become, Akim's memory of home was comforting. He hoped they had not been hurt too badly by the storm and floods. One of the other officers motioned to Akim.

"So, Darpheel, we haven't heard from you, do you want to head east with us to Sylhet or north with the colonel?"

"Neither." Darpheel answered. "My family isn't more than thirty kilometers up into the hills." He motioned behind him. "I am almost home."

"Then you're lucky. My family is, or was, in Chittagong. It might as well be on the moon from here, the way things are."

Quist, the sergeant who had accompanied Darpheel to the delta, spoke up, "At least they may be alright in Chittagong. My family is already dead in Dhaka." Everyone shook their head in agreement and condolence.

The group was roughly divided between the two camps, `go east' or `go north'. Soon discussions of options and intentions broke off and those who were not on guard looked for a place to sleep. Most crowded around the fire for warmth. Akim found that the black-brown coal retained enough of the day's heat to be comfortably warm at night, he carved himself a niche in the south side of the coal pile and lined it with a tarp. Akim had just settled in when a soft voice spoke out of the darkness.

"Sahib Darpheel? May I speak with you?" It was Latifa, the Major's widow.

"Yes. Ma'am. What can I do for you?" Akim answered.

Akim peered into the dark and he saw that the woman carried the younger of her two children. She stepped forward another step.

"Akim, can I call you Akim? " He grunted assent, she continued. "I have a favor to ask of you, a big favor."

"Yes?"

"I would like to come with you into the hills tomorrow. My parents' family was killed in Dhaka. With my husband dead, I have no more relatives I can turn to." She paused. "More travel will be hard on my children, and it is dangerous. If I can find a place to stay, close by, it would be best."

Thoughts raced through Akim's mind. What exactly was she asking? Akim knew a bit about her from talks at the office with her dead husband and others. She was from a wealthy merchant family in Dhaka. A beautiful woman, herself a university graduate, an engineer, she was not the type to ask what Akim thought he heard her asking.

Latifa spoke again, "I do not come empty handed. I have my dowery." She emphasized her words with a jangle of a leather bag heavy with coin, probably gold from the muted tinkle. She was indeed asking him to take over for her dead husband.

She spoke again, "I know it is a lot to ask, but it is not totally foolish, these are strange times we live in."

Strange, indeed, Akim thought. In the best of times back in Dhaka he could only dream of a beautiful and educated woman with a dowery. Now, in the oddest of circumstances, a beautiful, wealthy widow was thrusting herself on him. He thought for only a moment, then reached out to clasp her hand in response. Was it not a prime commandment of Allah that the widows and orphans should be cared for.

The next morning as Akim was helping Latifa gather her possessions from the truck for the hike into the hills, he was approached by Quist and the officer from Chittagong who had spoken the night before, they each carried a duffel, a Chinese rifle and several bandoleers of ammo.

"Captain Akim?" the officer, Lieutenant Mullahwi, asked.
Akim looked up. "Yes."

"The sergeant and I were talking things over. We are far from home and have no place better to go. Could your family use another pair of strong backs who are good shots? It is a good time to have a couple of trustworthy men at your side. And a couple of these." He patted his assault rifle.

Akim eyed the two men and thought for a moment. "It really isn't my decision. It is my father's and brothers' farm. But you are welcome to come with me to ask them. I, myself, would be glad to have you with me."

The young officer and sergeant looked at each other and then turned to Akim. "We can only try."

Both clasped Akim's outstretched hand and reached to help carry Latifa's belongings.

When everyone else who had made the choice of going north with Colonel Majlis or east in the other truck had loaded their things everyone gathered to say farewell to the others. Both groups embraced the new widow and wished her well. Darpheel was not openly congratulated on his good fortune, it was unspoken by his fellow soldiers.

Colonel Majlis wished them well and shook hands with Darpheel and Mullahwi. Quist saluted him and he returned the salute. Addressing these three and the others heading east, Majlis said, "It is unfortunate that events turned us from our duties. I wish there had been another way. Perhaps one day we can again serve the Bengali nation."

They all saluted Colonel Majlis as he joined his wife in the truck heading north.

As the trucks roared off to either side of the railroad bridge, Akim Darpheel, Latifa, two children and the two soldiers marched up the winding dirt road into the Garo Hills.

———

June - Newport Beach, California

The sharp point first touched Ian right behind his right ear. He had not heard anyone come up behind him, so he gave an involuntary start. But Ian knew what it was immediately. He waited silently for the other nine claws to sink into his throat. They did.

Presently the soft fingers attached to the manicured nails also grasped his throat and pulled his head back. He obliged by tilting his head back as far as he could.

The long blonde tresses whisked across his face before the upside down kiss found his lips. Cyndi had joined him as he sat on the balcony watching the heavy rain fall in sheets across the marina. She sat next to him on the hammock and he put his arm around her. They were a matched set, perfect for Newport Harbor; both had blonde hair, a good tan and wore yachting whites. They could have been a pair of models for a commercial.

"You want to talk about it?" she asked, sensing his somber mood.

"Not much to talk about. All this talk about the sea level rising has pretty much nixed the condominium deal. The attorneys who were going to do it for a share of the proceeds, backed out and want their fees up front. I can't find anybody else and this kind of thing needs lots of legal work. I can't say as I blame them. The real estate market for the whole coast took a big hit when the President started all this rising oceans crap. I talked to Henri and Claude today. They're unloading everything they can and recommended I do to."

"What do you think?"

"I might as well unload. We can barely keep the apartments rented above the break even point. And this weather isn't helping. Who said it never rains in Southern California. And when its not raining we've got the heat, even at the beaches. If the sea water really does come up everything will be a loss. The problem is, I sincerely doubt if anyone is stupid enough in this market with the cloud hanging over us to let me get my money out of this or even take over the note. Who wants to buy a future coral reef. Until this ocean level scare dies out, I just don't think we can get out of this. And I'm not sure how long we can last. We haven't had any new tenants in weeks and three more leases are up without renewal at the first of the month."

"What will happen if things don't turn around?"

"We could loose the complex in just a few months. No use putting good money after bad. The complex is in the corporation so we won't loose our shirts, but we would loose the property and all the seed capital we put in."

"And if the ocean does come up?"

"Well, I don't think they're really thinking its going to happen anytime soon, but when it does the whole restaurant and everything else in Newport would be gone. Not much you can do about that. Although I did see that article in the Times about building a dike to keep the ocean back."

"Anything we can do to protect ourselves, assuming we can't unload the properties?" She looked him in the eyes, hers were moist.

Ian had not wanted to scare her so he had not spoken of this before. He paused a moment before he spoke. "I set up accounts in that Nevada bank that gave us the deal on the Master Charge and Visa accounts. All of our personnel accounts, savings and the reserve for the business is there. We've still got that chunk of land in Nevada that Henri talked me into getting into. We should be able to keep the wolves away if everything goes to shit here in California, either with the ocean or the business. It will take the creditors a while to find assets in Nevada and it will be nice to have a bank that's above water if things go for the worst." He laughed a humorless laugh.

Lightning cracked to the south toward Laguna. The torrential downpour showed no signs of letting up. The couple turned inside and closed the french doors to the remnants of a tropical storm driven north into California by the El Nino effect.

They had no way of knowing that the storm was the harbinger of a string of uncommonly wet and uncommonly hot weather to pelt southern California with subtropical rain for months to come.

————

June - St. Petersburg, Russia

Cholera was a vile, disgusting disease that had no place in a modern industrialized city. Or so Dr. Sofia Litasku had thought before this spring. Somehow the dreaded scourge of undeveloped countries had invaded her city and was killing children in her hospital.

The first cases had come in the early spring. Thirteen students from the National Architectural Academy had come down with the symptoms in late April. Five of them had been too far gone when they received treatment to be saved. The rest had been saved only by the fortuitous actions of a school nurse at the academy. When the first student had come in vomiting, she had suspected drug use or bad liquor and had taken a blood sample which was sent to the central laboratory for processing. By the time the results were finished there were a dozen vomiting, diarrheic students being admitted to Sofia's internal medicine ward. The thirteenth had driven home to Novgorod to be with her parents before the cholera appeared and was hospitalized there.

When the lab report of vibrio choleris bacteria had come back and all twelve patients continued to vomit and soil themselves, along with having definite signs of dehydration, Dr. Litasku had started the intravenous injections and reported to the epidemiological service in Moscow. They had sent a team to investigate and assist. Their help was invaluable. They had brought with them the high potency tetracycline antibiotic that Dr. Litasku had asked for but never received. They also brought crates of injectable saline and lactose solutions.

The investigators from Moscow started to question the survivors as soon as they were able to talk. It did not take too long to glean the story of the drinking party along the Neva in April which had ended with a midnight plunge en masse by the naked students into the flood swollen waters on a dare. All thirteen students and four others who had not become sick had gone into the waters which obviously had been tainted by raw sewage during the floods.

But this had not ended the matter.

In early May, four residents of an industrial housing complex had come down with the disease. Two of them, a couple, had been found by neighboring tenants searching for the source of the foul odor in the apartment's hallway. The couple was found near death in a bed filled with vomit and feces. They did not survive.

The other two patients had come into Litasku's ward in time to save them. The investigative team from Moscow came again. This time they were not as lucky in finding the source of the contamination. These two patients lived in a different tenement than the two dead people. All of the water system for the area was on the same water main, but it checked out as clean. They could find no source for the fecal contamination of water, which was the usual source of cholera.

Now, a month later, the cholera had appeared again. Nine middle school students had come down with the disease. The children had been hit harder by the dehydration and seven had died. Then, two weeks later another sixteen children at another school had started vomiting and diarrhea. In the interim between outbreaks at schools, seven more adults from various occupations and residences came down with cholera. The difference in incubation period and location meant that they must find another source of the bacteria.

The newest outbreak was two days old. Nine of the sixteen children and two of the adults were in Sofia Litasku's hospital and under her care. Four children were near death and others could slip into the deadly dehydration spiral any time. The team from Moscow was drawing a blank on the source of the disease. The water system at school appeared pure and the children although from the same section of St. Petersburg, came from several different neighborhoods. The kitchen at school was searched and found to be spotless, more so than normal, the headmaster was a retired Navy officer who was a fanatic for cleanliness.

It was the cholera and the wave of hysteria that was moving across the city that brought Sofia Litasku to discuss the problem with her husband. If the public health department detectives had failed to find a cause of the epidemic, then perhaps her husband could get his own subordinates on the police force to find the source of the disgusting death that was striking down the children.

Sofia and Istvan sat at the dinner table. Nikolasha was reading a story to his granddaughter in her bedroom.

"This is the whole file from the investigating team?"

"It is what they gave the hospital. You might be able to get more, if you asked them."

Sofia thumbed through the stack of documents until she found a map, folded in two. "Here it is. This shows the spread of the cases. They are everywhere. No pattern."

Istvan considered the map. There did not appear to be a pattern. Discounting the first academy students for whom a cause was apparently known, the crosses on the map covered the whole eastern and southern sections of the city. The phone rang and Sofia went to answer it.

When she returned Istvan asked, "And the water was tested more than once?"

"Every hour for both of the schools and the tenement building."

"Well, perhaps we will have a look at this. It may not be our specialty, but I think that the militsia should investigate the deaths of fourteen citizens."

"Fifteen. That was the hospital. Another child!" Sofia choked back tears. She could not help but think of her own daughter being in the place of the children at the hospital.

Commander Litasku asked Detective Kolya Ivanovich Tsverchenko to assist him on the investigation. Litasku's managerial duties would not let him put his full time into the case and Tsverchenko was his brightest man, fresh from a big success in breaking the back of a counterfeiting scheme.

After studying the file they went to the first school. They interviewed the headmaster, the janitor, the cook and several teachers. They even talked to children on the playground. They found nothing other than that reported by the health officials.

The apartment buildings were likewise a dead end. The buildings were widely separated and the water was clean, apparently, each of the huge concrete slab tenements had several thousand still healthy residents. The work sites of the first four adults stricken were also in separate areas and apart from the new adult cases.

They went to the school with the recent cases. Litasku went to interview the headmaster while Tsverchenko went to the kitchen. It was lunchtime and Tsverchenko had to wait for the head cook to speak to him. The detective had just started to ask questions of the cook when Litasku walked into the kitchen.

"And where do you get your food supplies?" Tsverchenko was asking.

"The Education Ministry has contracts for most everything. They have a schedule and they send us whatever is on the schedule."

"What do you mean by 'most everything'?"

"Well, the ministry is in Moscow. They can only supply the non-perishables; canned goods, canned fruits, cured meats, flour, sugar and that kind of thing. If we need anything else, the headmaster has an account to pay for it, but it is small, we have to make do with the commodities from Moscow for everything but special occasions."

"And what are the special occasions?"

"Well, for instance on May Day we had pastries for the children?"

"Anything since May Day?"

"No, nothing."

Litasku exchanged looks of frustration with Tsverchenko. They were interrupted by the cook adding, "Except..."

Litasku spoke, "Yes, except what?"

"The republic laws require each school to serve fresh vegetables at least once each week." He looked at Litasku; a hint of fear in his eyes. "We buy the vegetables from the headmasters fund, from the central farmer's market."

"And what did you buy in the last two weeks?"

"Cabbages. Fresh cabbages. The first crop of spring cabbage is in. We have had them for three weeks. All of the schools buy them. They are quite reasonably priced."

"But you cook the cabbages?"

"No, the law says they must be fresh for the vitamins. We make cabbage salad, coleslaw."

It had not taken much work to confirm that the other school had used the farmer's market for its produce.

For once the Russian propensity for administrative attention to detail paid off. Both headmasters had required their cooks to produce receipts for the goods purchased on the schools' cash accounts. The cabbage seller was the same, a farmers' cooperative which sold the produce of local farms at the central market.

Litasku and Tsverchenko went together to the market, a huge indoor arcade with long lines of shops selling many foodstuffs, fruits, vegetables, meats and whatever else was available. Back in the '90's, with the advent of the market economy free markets like this had sprung up throughout the country.

A few questions located the farmers' cooperative. They had a rather large stand occupying several stalls near the side entrance. Without asking any questions of them, the two policemen watched the cooperative's stalls. They were doing a brisk business.

As the stalls emptied of vegetables, the old women working at the stand refilled them from large crates behind the stand. The crates were lined with plastic liners. When the old women took the cabbages, lettuce, beets or whatever out of the crates they shook them to get rid of the excess water. The vegetables had obviously been washed before packing and kept damp to keep them fresh. Litasku and Tsverchenko had seen enough.

"Yes, what do you want?" The old woman did not look up when Litasku came up.

"Who is your manager?"

The woman looked up and saw the Militsia Commander and another officer. She yelled behind her, "Yuri, come here?"

Yuri, a wizened old farmer, came out of the alcove behind the stand. He looked at the officers, smiled and said, "Yes, May I help you?"

Tsverchneko flipped open his identification wallet to show his militsia credentials. Litasku did not bother, instead he spoke. "We have some questions. Do you wash your produce?"

The now nervous Yuri spoke as he hurried back to a crate of cabbage, "Yes, yes, yes, everything is washed and thoroughly, too." He grabbed and shook a head of cabbage. "See, it is still wet."

"And where do you wash them?"

"At the main farm. Our warehouse. Much of our produce this time of year comes from farther south, we wash it and sort it at the old collective farm. It is just outside of town on the road to Novgorod."

"What water do you use to wash them?"

"Why, our farm water. We have a good well and a tiled cistern. Why do you ask?"

"And tell me. Did your farm get flooded this spring?"

"Certainly. We were lucky to get our fields dry enough to plant in April."

A flooded cistern.

Litasku and Tsverchenko looked at each other. Litasku went to phone the hospital while Tsverchenko informed Yuri of his farm's predicament.

The two policemen had solved the riddle and nightmare of the St. Petersburg cholera deaths. The epidemiological team from Moscow would make reports to the World Health Organization as cholera was one of the virulent plagues which the nations of the world took note of. It was feared that such reports could become increasingly common as floods and rising sea levels came to more countries, especially in the developed nations who had never had need to worry about the quality of their water. Cholera, endemic in parts of Asia, Africa and South America which relied on impure sources of water, could become a familiar occurrence in modern metropolitan areas with complex and intermingled sewage and water systems, massive sewage overburden, ocean discharge of wastes, and storm and flood drainage systems designed for a different climate and ocean level than these areas could face in the future.

July - Southern Connecticut

It was good to be home. They should do it more often.

Andy and Molly Knowles always enjoyed their rare chance to come to the house in Connecticut. The only times he had enough time to get here to the house were the infrequent congressional breaks and these had to be carefully planned to coincide with pre-scheduled vacations from Molly's law firm. There was an ironic twist to finally having the wherewithal to own a beautiful place like this, but having lost the time to enjoy it. Andy turned into the lane that served both their house and two neighboring farms. The rough ride up the lane roused Molly from her nap. She shook her brown pageboy hair out of her face and looked at her watch.

"You know, honey, this will be the first time we have been alone on vacation in years. I can't remember any two person vacations since Chad was born." Molly said after a long yawn.

Andy looked over to her, thought for a moment with pursed brow and then nodded. It was the first thing she had said since Essex. She had slept most of the way down from Hartford except for the obligatory stop in Essex for food and supplies. It was difficult maintaining two households, in rural Connecticut and in suburban Bethesda, especially when you only got to visit the more comfortable of the two homes four or five times a year.

They always stopped in Essex, not only for supplies, but to check in with the real estate firm that watched over the house in their absence. Once, during Andy's presidential campaign the house had been ransacked by vandals during the Knowles' time away, causing Andy's Secret Service contingent no end of worry. Now, both the management company and the local constable kept an eye on the property.

Sandy was off to California for a summer college program in ecology and Chad had just reported to flight school in Texas after graduating from the Air Force Academy in May. Thinking of this, Andy realized that this would indeed be his first vacation alone with Molly since Chad had been born. Perhaps the situation was not so unfortunate after all. Also it would not be totally alone, not with two different sets of guests coming for visits and Andy had several appointments around the state during the next few weeks.

The house was a two story white farm house, actually three with the gabled attic. With a matching barn the homestead was postcard perfect, thanks to Molly's determination to make her mark on the property. They had purchased the big old house during Andy's first term as Congressman. The choice had been completely Molly's. The traditional Old New England style house was as near as she could get to her family home outside of New Haven, but still remain in Andy's congressional district which centered on his boyhood home in Norwich. The house was also a compromise, not only in location, but in style and size, from Andy's boyhood home. Molly had not felt comfortable in the modest frame house Andy had inherited on his father's death in a working class Norwich neighborhood. Molly had grown up on what Andy always called an "estate" in Branford and this roomy house overlooking the Connecticut River was more of what Molly thought a Congressman, now Senator, should be living in.

"Everything looks OK," Andy surmised as they pulled to a stop.

"Hhmmff," was Molly's only response, she was still groggy from her nap.

Andy unlocked the two big locking hasps on the kitchen door and as Molly carried the first load of groceries into the kitchen, Andy went around and unlocked the closed wooden shutters on the ground floor windows. It would not be long before the big, old house would be home again.

"Now that you're done puttering, I couldn't interest you in a walk down to the boathouse, could I?" Andy asked Molly.

"Just as long as the walk isn't down a primrose path." Molly chuckled back.

It was an inside joke; one of Andy's presidential campaign opponents once had called Andy's economic program "Primrose path economics" and the appellation had haunted Andy throughout the primaries. It had come up again the day before in Hartford. Andy had made a speech on environmental issues to a Fourth of July crowd which included a several laid off factory workers carrying placards labelling Knowles' environmental activism as "another primrose path." The Hartford media had played up the jobs versus environment issue in response.

Andy sneered a response and playfully pushed his wife to the door. They were both dressed in the blue jeans and t-shirts that always felt so comfortable after the business suits and stiff shirts and blouses of Washington, D.C. Outside, the weather was beautiful, the late afternoon sun bringing the best in color out of the almost totally green countryside.

"I can't recall it ever being so lush," Molly said as she carefully avoided patches of mud in the track down to the river. Andy nodded agreement, his attention on his surroundings, also.

Their house was on a hillock overlooking the lower Connecticut River. They had about forty acres of land. The property had been a bit expensive for a freshman congressman, but Molly's trust fund had made the difference. The argument over using the trust fund to buy the family house had been one of their few marital battles, with Molly accusing Andy of being chauvinistic about her input into the marriage and Andy arguing that they should live according to their immediate means not on their inheritance. When Molly pointed out that by selling his father's house in Norwich he was using his own inheritance Andy finally relented. After that experience Molly had learned how to be discrete in how she used her parents' wealth for the benefit of Andy and their family. As Andy became more involved in Washington politics, Molly took over more of the family financial affairs. The argument over using her money had long since passed. Three years earlier she had bought him the cabin cruiser that was housed in the boathouse they now went to check on. He had never commented on the source of the money for the expensive boat.

The main pasture area of the farm was overgrown, making the walk difficult and the track was even more muddy the lower down the hill they walked.

"What happened to the farmer that came in and cut this down a couple times a year," Andy commented to Molly.

"Oh, I thought I told you. He died and the management company can't find anyone that wants to come way out here to cut it."

"Oh, well," Andy shrugged in resignation.

"Andy, I have some good news I have been waiting to tell you."

Andy stopped in his tracks and looked at her. "You're not, ... you know... ahh...."

Molly laughed at Andy's flustered look, " No, not that. Thank God. That would be all I'd need right now. We've done our bit toward procreation, dear. No, the partners called me in yesterday and asked if I wanted to be taken off 'of counsel' status and made a named partner."

Andy turned to Molly, "That's wonderful, I'm proud of you." He hugged her.

Still in his arms Molly said, "If you don't see any problems, as of next month it will be Jennings, DuCroix, Ellerson and Knowles."

For just a split second Andy's hug relaxed as he thought of what she had said. Then he pulled her out to arms length to look in her eyes. She smiled. He smiled. They hugged again. The "and Knowles" was somewhat of a milestone in their marriage. After leaving law school together as a young married couple, Molly had kept her maiden name, out of honor to her prominent family, the Hobart's, and the feminist attitude which she espoused.

With the advent of children and Andy's political career she had reluctantly modified her professional surname to the hyphenated 'Hobart-Knowles". Now, with this bit of news, Molly was giving Andy a gift that she knew he appreciated, Molly, the law firm partner, was calling herself Molly Knowles, not Molly Hobart-Knowles, for the first time. Andy acknowledged it with as warm a hug and kiss as he could give. Molly responded. It was only after several moments that the couple realized that they had been standing in an open field on a summer afternoon, lost in an embrace. They continued on down to the boathouse, there would be time for each other later.

The area around the boathouse was even more soggy and overgrown than the rest of the field. A line of dried algae could be seen about two feet up the side of the wall. The boathouse was one an estuary somewhat away from the main channel of the river. But it obviously still had a tidal wash. The high tides that had hit the rest of the eastern seaboard had not spared southern Connecticut.

Andy opened the door to the boathouse and carefully felt his way down the inside wall on the plank walkway. The big Bayliner nearly filled the old boathouse, making it necessary to turn sideways to clear the widest part of the bow. The planks were slippery with algae. The high water had, of course, come inside the boathouse, too.

Andy unlatched the boat door and Molly pulled it open by the rope that served as a handle outside. A moment later she pulled the other rope and the other half of the door swung out.

Andy was wiping the instrument console when Molly made her way down the slippery planks.

"She's in pretty good shape for not having been touched in over a year," Andy noted with satisfaction.

Molly looked quizzically at him and said, "Chad took the boat out last month when he came up for his things after he graduated. He wrote about it in his letter."

"I guess I'm not keeping in touch very well. I didn't even know he wrote." Andy shrugged it off again. "Let's see if she starts."

Andy pulled out the choke and switched on the ignition. The starter growled and spun. After two growls it coughed to life. When the engine roared to life the electronics came on, including the radio, set to the coastal advisory channel. Andy started to reach and turn it off, but waited when he heard the broadcast.

"...watch will be in effect for all areas in New England and the mid-Atlantic states. All vessels and small craft in the hurricane watch area are advised to return to port or leave the affected area. All shore facilities are advised to make preparations for heavy weather. Should the watch be upgraded to a warning, additional instruction for evacuation will be forthcoming." Andy switched the engine off.

"Well, there goes the casual boat trip up to Norwich idea." Andy said in a sarcastic manner.

"I thought the weather report back in D.C. said Hurricane Clark was going out toward Bermuda?"

"They just have a watch on for it, if it were likely to be coming they would issue a warning. They had a watch for Virginia and Maryland on south yesterday, but the storm kept well out to sea. Now it has followed us up the coast."

"Think we should stay here?"

"We're at least ten miles inland from Long Island Sound. Besides, that old house has weathered worse winter Nor'easters than anything that's going to hit us this summer. Besides it's only a watch, if anything changes we will have time to reconsider. We'll watch it on the TV and if it looks bad we can get out then."

Now it was Molly's turn to shrug. He was probably right. They closed up the boathouse and walked hand in hand back up the path to the house. The sky to the south over Long Island Sound was brilliant blue with lazy white puffs of clouds. It certainly did not look like hurricane weather.

———

July - Boone, Iowa

The small puffs of dust stirred up by his boots in the furrows drifted back to earth without any wind to blow them off. Thank God for that. If there were wind to speed the evaporation it would be even worse. Worse? Worse than what? Did total disaster come in degrees?

Only in the shallow depression in the southwest quarter section did the corn come anywhere near his knees. "Knee high by the Fourth," was the old saying, and it should have held true for Jared Tolliver's corn. But this year it would not be.

The last of the spring rains had come in the first week of April and Jared and his neighbors had soon thereafter given thanks that they would not have another horribly wet spring that kept them from planting again, as it had for three years running. The fields dried out considerably and they were able to buy the seed to plant and get the fields planted in plenty of time to have a good yield in the fall.

As he had predicted, the west quarter section along the county road had dried out first. He had gotten it planted by mid April. The weather dried out and he was able to get the other two quarters in by the end of April. He had planted a high yield hybrid in the first field. It would have a maturity of 120 days plus, so by mid-August he should have a good crop. The southeast quarter got soybeans and the wettest quarter, the southwest, was planted in 90 day hybrid corn.

The break that the weather gave the Iowa farmers in April turned out to be too good to be true. They had been accustomed to the wet springs for three years running. And so, when the skies dried up in April they had rejoiced. Prematurely.

The glory of Iowa agriculture was the combination of fantastic soil, coupled with the hot sun and occasional rainstorms of May, June and July which combined to give a perfect growing season for the breadbasket crops Iowa was famous for. In the previous years' crop failures the rain had come too early and they had not been able to plant. This year the rain let up in April for them to plant, but did not come back. No amount of soil moisture from April could carry crops through to July without rain. Everything was drying up.

As Jared turned back toward the house, he saw the brown Cherokee 4x4 of the county extension agent heading north on the road. It turned into Jared's drive. Jared picked up his pace toward the house.

When he got to the yard, he saw that two men had exited the vehicle and were waiting for him. One was Oliver Mortenson, the extension agent. He did not recognize the other.

"Goo' Morning, Jared," Mortenson said as he shook Jared's hand.

Mortenson turned to the other man and said, "This is Jared Tolliver. Jared, this is Doctor Sean Reid, from the Ag Research Lab at Ames."

The two men shook hands. Mortenson continued talking, "Doctor Reid is making the rounds with me. Trying to get some soil tests done for the government. You wouldn't mind if he took some samples would you?"

"Hell, no. This worthless mess might as well be of some use to somebody."

Reid Spoke, "Yeh, we saw it coming in. Pretty bad. It's like that all the way across the state. We have a wide band from Pottowatamie County to Dubuque that hasn't gotten a drop of rain since the first of April. That's what I'm checking on; moisture content of the soil after a rainless stretch like this.

"A couple scientists with the federal government have an idea that the increase in carbon dioxide in the air is drying the soil out at an accelerated rate. They want each state to give them data. So, if you don't mind?"

"Go right ahead. Can I help with anything?"

Walking to the back of the vehicle, Reid said, "As a matter of fact, yes. We need a plug of soil from the middle of each quarter quarter. I understand you farm three quarters. that'll be twelve plugs. So. if you two could help me carry the plug pullers out." He reached in the tailgate window and handed Mortenson and Jared each four hammer-shaped tools consisting of a long tube and a handle. Reid took four more of them and a clipboard. The three men headed for the fields.

Jared led them to the dry northwest field first. As Reid pushed the tube into the ground, Jared spoke. "What do you two think of all the environmental news? Did you see The Register last week? Those maps?"

Mortenson answered, "Yeh. Scarier than hell, if you ask me. At first, I was thankful we were in Iowa and not on the coast. And then we started getting crop reports and drought conditions set in. Looks like we might be in for it."

"It sure made crop futures go through the roof. Too bad I'm apparently not going to be able to take advantage of the markets." Jared added.

Reid spoke, "I went through the whole packet USDA put out on the climate changes. If you believe what they say, we can count on being more like Arkansas or Texas climate-wise, than what we know of as Iowa. And that is if we're lucky. We might be more like New Mexico or Arizona if the Jet streams and upper air systems don't take shape just right."

No one talked as they walked south to the one somewhat green quarter. Reid took his samples.

Jared broke the silence as they turned toward the last quarter, "What do they grow in Texas anyway?"

Reid answered, "Lots of cotton, rice in the wet parts. Where its irrigated, they grow most anything. They have a growing season of anywhere from 180 days to 300 days. Of course, where its drier they turn to cattle and livestock feedgrains."

"You really think that our climate is switching to that?"

"Damned if I know. But, our record sure is looking that way. The latest killing frost in five years is March 13th. And we've had more precipitation from March 1st to May 15th for the last four years than we have had in the rest of the year combined. The charts for corn production have gone down for the last six years and soybeans have gone down for three. That sure doesn't sound like the Iowa I know."

Jared shook his head, half in disbelief, half in resignation, as he watched the scientist push the soil sampler into the dust between stunted soybean plants. Rice and cotton, huh? What did he know about rice and cotton?

———

July - Southern Connecticut

The state trooper who drove the governor's car had it running with passenger doors open when they came out. He wore a dark blazer with his badge on the pocket rather than the typical trooper's uniform. He waited by the open door.

Andy and the Governor Drexel, and Molly and Mrs. Drexel came down the porch steps, two by two. Each pair conversed to each other. To an outside observer Morgan and Judith Drexel could have been Andy and Molly's parents, such was the difference in apparent age and appearance. The governor and his wife were in business clothes and the senator and his wife were considerably more casual in their dress.

"Well, Andy, I want to thank you for having us over. I'm glad we still can talk man to man on the important things."

"Your most welcome. I'm glad for opportunities like these to keep in touch with what's happening here at home," Andy Knowles replied.

"...and, Molly, you simply must stop in the next time your in Hartford. You won't recognize the place. I've redone everything for Morgan's re-election." The governor's wife stopped talking a moment to pat her side for her purse.

"We'll make a point of doing that, Judy." Molly replied as they came to the car. The trooper closed the front door for the young campaign staff member who had accompanied the governor and waited for the foursome to finish talking.

A few light raindrops fell from the hazy, overcast sky. The governor looked skyward and commented, "Looks like a bit of the hurricane's weather is skimming us."

"Any word on what we can expect."

"The latest I heard was that it was heading east out to sea. Nothing for any mainland areas to worry about. But Cape Cod, Nantucket and Martha's Vineyard may get soaked and buffeted a bit."

"Good news for us!" Andy replied, Molly nodded and hmmmed.

"Us, too. Next stop is a big open air election rally with the shipyard and waterfront workers in New Haven. Wouldn't want to weather to mess it up." The Governor said as he and his wife turned to shake hands with the Knowles.

To Molly, Governor Drexel said, "Thank you for a wonderful lunch, I wish the campaign trail had more breaks like this."

Everyone else said their goodbyes and the governor's black sedan finally pulled away and went down the lane. Andy and Molly stood on the porch and waved to the departing car.

"Thank God that's over. If she had mentioned one more charitable venture I could work on in my `spare time' like she does, I would have killed her," Molly said.

Andy laughed. "I guess you just don't have what it takes for the patrician life. That's why you felt the need to marry below your station and become a working girl."

It was Molly's turn to smile. "I wouldn't call being a Senator's wife and almost First Lady `below my station'." She turned to follow him into the house. She slid her arm around his waist as they climbed the stairs.

"Almost doesn't count. I didn't get you the White House."

"No, not yet."

Andy was helping Molly clear the dishes from the lunch.

"I've gotten very lazy with the maid to do this in Bethesda." Molly said as she clamped the door of the dish washer closed and turned it on. "At least that's one visit down and one to go. I guess I can be thankful that you didn't invite the Drexel's to stay over. What on earth got into you to invite someone I don't even know to spend the night."

"It just worked out that if I was going to get to talk to Brady without driving in to New London to meet him after his conference, that it would have to be here. And since it was just him and his wife, it just seemed best for them to stay over before going to the airport the next day. Besides, Brady is a good man and I wanted to get to know him better."

"So what's the scoop on him anyway and do you know anything about his wife?"

"I suspected that you had slept through my whole discussion of this on the way down from Hartford and this confirms it."

"You politicians briefing an issue always puts me to sleep. Go on." She sat a cup of coffee on the kitchen table and he sat down.

"Well, you know most of what went on in Boulder and subsequently. Brady, Ken is his first name, is former military, Air Force I think. Taught at the university in Boulder for a while, doctorate in meteorology. He's a little bit younger than us." Andy sipped his coffee and continued. "I have never met her; Virginia, she goes by Ginger. Derek's background report says she is a commercial artist in Boulder and teaches part-time at the university, I guess that's where they met.

"Brady is over in New London today for a conference at the Coast Guard Academy on the maritime effects of the global warming. Ken Brady thought enough of me to give me all of the Gaia information when he needed help. I respect him and thought I would take the opportunity to find out more about him and what has gone on in his work since March when this whole thing broke."

Molly nodded,"OK, it may not be so bad. But I swear, if this Ginger babe does charity work and has a need to proselytize me and my `spare time' for her worthy cause, I may not be responsible for my actions."

"Fair enough."

———

Andy braced himself against the stiff wind and spattering rain as he went to the car. Ken Brady got out to meet him.

"Welcome. Hope you didn't have any trouble finding us," Andy said as they shook hands.

"No not at all"

Andy saw that his wife had followed him out and was greeting Brady's wife on the passenger side of the car.

"Hi, I'm Molly Knowles."

"Glad to meet you, I'm Ginger, Ginger Brady." They shook hands. Ken and Andy both introduced their wives to the other and all acknowledged the introduction with a nod from either side of the car. Andy took a moment to compare the two women and their contrasting looks. Although both were very attractive, they were on opposite ends of the scale. Molly's curtly professional brown hair and patrician looks versus Ginger's billowing hair and earthy blonde beauty.

When Ginger started to open the back door to get the baggage, Andy spoke up.

"You can go on inside, we'll get that."

"Ahh, chivalry lives," Molly said. She hunched her shoulders in the cold wind and led Ginger up the steps.

Ken and Andy carried the bags into the foyer and sat them down. They both took off jackets spattered with rain drops.

"Damn, getting a bit nippy for July. You didn't bring the hurricane in with you, did you?"

"Not hardly. Wind is still in the northeast, if it starts switching around to the south, then you can start thinking hurricane. Last I heard some of the lee edge might hit outer Cape Cod though."

"I forgot for a minute that weather is your stock in trade. For the rest of us it is a bit more mysterious."

"Nah, nothing mysterious, it's a science. Now politics, that's another story, that's mysterious." Both men laughed.

They were interrupted by the sound of Molly pushing the sliding doors to the den open. "Andy's got a good fire going. Why don't you all come in and get warm."

They did.

Molly and Ginger had left to show Ginger where to change from her travel clothes into something for dinner. Andy and Ken sat before the now roaring fire.

"So, how did the conference go? Learn anything new," Andy asked Brady.

"I am always amazed when we think we have a handle on something and then something new pops out. We, at GAIA, thought we had the ocean level data tied up tight. Between La Jolla and Boulder we had the projections for how much the thermal expansion would raise sea level for every additional degree of global warming. Then I come to this conference and a British team from the University of Liverpool shows the results of their study of variations in ocean rise at various latitudes. Their data shows pretty irrefutable proof that for every unit of sea level rise at the Arctic Circle there is a two unit rise at the Equator.

"We had always assumed that the projected two to four foot rise from thermal expansion would be uniform. Ocean levels are a somewhat static reference point and in human history we have never had any way to or reason to judge whether water in the oceans flows more readily to one point than another. It never occurred to us to have the dynamic of the centrifugal force of the Earth's rotation reacting to different temperature water as a factor for sea level rise."

"Sounds like good news for northern Europe and America and bad news for the tropics," Andy conjectured.

"Yes, but that's not all. The Russians from the Moscow Institute for Geophysics not only agreed with the Brits, but added one step to the process, by figuring the Coriolis Effect into the equation. They figure that the same Coriolis Effect that causes water going down drains to spin counter-clockwise in the northern hemisphere and the reverse in the south will cause a magnification of the tidal surge in some bodies of water. The formula is pretty complicated and still hasn't been confirmed a hundred percent in the real world, but they expect the eastern shores of bodies of water facing west in the northern hemisphere and the western shores of water bodies in the southern hemisphere will have pronounced increases in tidal levels. For them it answers the mess they have at St. Petersburg and for us it explains the salt water intrusion problem out in the Sacramento River delta off of San Francisco Bay in California and Puget Sou nd."

Andy considered this for a moment and asked, "So what is the best guess on the sea level rise we can expect."

"It all depends on how long the temperature increase continues. At current rates the annual increase could be something like a few inches per year for the east and west coasts of the U.S. and up to a half foot for places like Lagos, Nigeria and Singapore. That's assuming we don't have any significant input from melting glaciers. If the temperature increase continues down to deeper ocean depths it could go right on for years."

"Damn," Andy said as he shook his head, he started to say something else, but Ken continued.

"As I said when we first talked out in Boulder, the sea level increase may be the first physical manifestation we see, but in the long run, the climate change may be the most profound effect of the warming. The Scandinavians at the conference are really worried that the increase in global temperature could alter the ocean currents. A change of a few degrees in certain currents or a few percentage points in salinity could cause catastrophic changes in certain actions of some convergence zones."

"Which means?"

"They're afraid that a warming of the Arctic coupled with more fresh water melting off of the Greenland ice packs could cause the Gulf Stream to cycle itself farther to the south. If that happens, Sweden and Norway will make Siberia look warm and the British Isles will have Labrador's climate. Portugal could get the rain Ireland usually gets and the deserts of Morocco and Mauritania could have the same climate that the farming region of Bordeaux, France now has. We're talking weather chaos not seen since the last Ice Age. We have already seen a similar effect in the El Nino phenomenon in the Pacific, but if it sabotages the Gulf Stream the highly populated food producing regions throughout Europe as far east as the Ukraine could be devastated. And the evidence suggests that it could happen in as little as one season, whenever the balance swings too far for the deep ocean currents to let the Gulf Stream push into the North Atlantic."

The senator had not yet responded when they were interrupted by their wives at the door. Molly spoke, "If you gentlemen could break away from your global warming conference, dinner is waiting."

The men followed the women to the dining room where Molly adeptly focused the dinner conversation on personal and family matters rather than the talk of weather, climate and politics that Andy and Ken had the urge to discuss. Ginger caught the drift of Molly's questions and the two women soon had their husbands exchanging life histories and anecdotes rather than the somber news of potential disaster which had brought them together.

By the time dinner was finished, the wind had picked up outside, rattling shutters and rustling trees.

———

July - Christchurch, New Zealand

The sound of the clock radio slowly intruded into Tim Wesley's consciousness. Dorothy rolled over next to him and covered her head in the pillow. He had to think for a minute about why he had set the alarm so early; it was still pitch black outside. What day was it? Did he have to go to work?

It was only the weather report on the radio that brought him to full consciousness. He had heard a weather report on the telly last night and the weather for Christchurch and rest of the South Island was forecast as excellent today. The storm systems that had brought continuous blizzards and overcast winter skies to much of New Zealand for weeks had finally given way to a high pressure system. Today would be the first good skiing weather of July. Clear, relatively warm and sunny and with two weeks build-up of fresh snow. Of course. He had set the alarm to get the family up for skiing.

"Come on," he said as he nudged Dorothy in the buttocks. Then thinking better of it, he followed with a kiss to an exposed shoulder.

Tim hopped out of bed and headed for the bathroom, but reconsidered. He had to wake up the kids or they would never get going. He went first to the boys' rooms.

"Everybody up and at 'em," he shouted as he banged on doors and flipped light switches up. He started to do the same with Alice's room and then he remembered she had called her college friend, Hillary, late last night to go with them to the slopes today. He would give the young women a bit more courtesy.

"Good morning, ladies. Mount Cook awaits," he announced as he knocked on their door.

Tim heard sufficient moans and rustling throughout the bedrooms to think that the wake-up was in progress. He went to his bath room and started the shower. By the time he came out Dorothy had his cold weather underwear and ski pants laid out on the foot of the bed. She was a dear, it was too bad he had never been able to get Dorothy to try skiing herself.

———

As usual for these ski trips, the Range Rover was packed full. The three boys had piled into the far back seat with Alice and Hillary in the middle seat. Dorothy was beside Tim in the passenger seat, already busy at work on the afghan she would spend her day knitting in the ski lodge, when she was not reading.

Just outside of Ashburton, about half way across the Canterbury Plains, the sun had finally made its appearance. It had been a glorious sunrise out over the Pacific. A few of the last storm clouds were still in the far off eastern sky and they blazed with the purples and oranges of dawn. Even the boys had quieted down for a moment to watch the kaleidoscope that was sunrise that morning. Tim Wesley could hardly ever remember a sunrise so brilliant and beautiful.

They pulled into the ski area above MacKenzie well ahead of the crowd Tim expected on this Saturday. The ski buses from Christchurch would not get in until mid-morning. The Southern Alps were majestic in the morning light. By the time Tim had returned from the shack with the lift tickets everyone was suited up, boots on and ready to go. Except, of course, for Dorothy who was carrying her knitting bag and book up the stairs to the lodge where she would spend the day.

Jerome and Peter had matching blue snowsuits on and with the matching headbands and goggles you could not tell them apart. Tim snapped the lift tickets onto their zippers and they were off without a word.

Tim, Jr. took his ticket, as did Alice and Hillary. Neither girl had ski goggles on, wearing only sunglasses.

"You girls sure you don't want your goggles? You'll loose those things the first fall you have," Tim asked.

"Ah, Dad. You know how I hate to look like a panda for weeks after a ski trip," Alice responded.

Tim, Jr. cut in, "Pandas have dark circles around their eyes, not white."

"Shut up, creep," his older sister retorted.

"Alright, alright. Just be careful. All of you." Tim's last words were not heard in the scramble of the youngsters toward the lifts.

Tim rode with his son up the main, central chair lift. The girls were two chairs in front of them. The twins were nowhere to be seen. The ride up the mountain was breathtaking. Below was trackless, fresh powder; brilliant white in the morning light. The weather was not too warm and not too cold, perfect for skiing. The sun was warm on the back of his neck, the westerly breeze brisk on his cheeks.

At the top of the first run Tim, Jr. did not waste any time schussing quickly downhill away from his father. The girls took a side trail to the south, so Tim Wesley was left on his own. He chose the main slope and its gentle mogul field. He took his time and enjoyed himself, it was a great day for skiing.

The first run had taken about an hour, counting the lift ride. At the bottom, Tim did not see any of the family and not wanting to stop and go into the lodge with Dorothy yet, he decided to take the northern lift up for another run.

He rode alone in the chair for the long trip up to the northern bowl. It was a two lift run, where you take one lift up the mountain, then ski into the bowl on the far side. From the bottom of the bowl you take another lift up to the upper reaches of the ski slope and ski down the outermost trails to the main lodge area.

Tim took off his mittens and carefully put them under his leg so they would not fall off the lift. He pulled his goggles up over his stocking hat to rub his eyes. He winced at the brightness of the snow-reflected sunlight.

At the first touch of his hands to his cheeks he knew of his mistake. He had not put on enough sunblock. The crinkly pain of his cheeks told him he would have even more trouble before he finished the next two runs and returned to the lodge were Dorothy would have the sunblock in her bag.

Feeling the same sunburn on his exposed neck, Tim zipped the parka to the top. He carefully placed the goggles exactly where they had been to avoid the burnt areas. He replaced the mittens and exited the lift.

He skied down the open treeless slopes of the northern bowl without incident. Now he could feel the pain in his cheeks whenever he moved his face. Before boarding the chair back to the top of the mountain, he took a moment to pull his stocking cap all of the way down over his forehead.

There were very few people boarding the chair out of the bowl. The lift operator was inside the lift house, so Tim boarded the lift himself, jumping into the tracks and bending his knees to catch the chair just right. Without the lift operator to restrain the chair at the last moment it hit him hard behind the knees, but it was not a problem, he was getting anxious to get to the lodge.

On the way up the lift he pulled his scarf higher around his face. He was bundled up as well as could be. The lift run was to the southwest so he had the sun at his back, but the light reflected from the slope before him was almost as bad as the direct sun.

The trails from the top of the mountain wound down around and through the poles carrying the main lift he had ridden on first run. As he looked up he saw that it was nearly empty. Only a few chairs had passengers and they appeared to be wearing the terra cota colored jumpsuits of the Ski Patrol members.

Tim caught a ski edge in an unseen rut under the powder, missed a turn on one of the trails and took a dive into a bank of powder. He was not hurt. His right safety binding had popped the ski from his boot. He was totally covered in snow.

In what was an automatic action for a skier, he flipped the goggles up on his forehead and wiped the snow from his face.

And winced in pain.

How had he gotten such a sunburn so quickly?

Tim carefully wiped off the rest of the snow, rolled on his side and then reset his boot in the binding. Getting to his feet, he carefully replaced hat, goggles and scarf.

He carefully covered the final trail which intersected the main slope he had come down early that morning some two hours before. He was amazed at how few people were on the slope, far fewer than the early run he had made.

About three hundred meters from the bottom he started hearing a loudspeaker blaring out a message of some kind. As he came closer he finally heard it clearly.

"Attention all skiers. The slopes and lifts are closed. All skiers are directed to report to the main lodge for instructions. Attention all skiers...."

What was going on?

The lodge area was an incredible sight. As skiers came off the mountain they were immediately herded into either the lodge or one of the other buildings by Ski Patrol members and policemen. Both Ski Patrol and police had scarves and/or ski caps covering their face and goggles on. As he came to a stop and stepped out of his skis, Tim noticed the woman who had come to the ski racks beside him was in tears. And her tears rolled down blistered, cracked cheeks.

A policeman tried to usher him into the line to enter the ski repair shop and equipment sales office, but Tim protested, "My family!," and pointed toward the lodge where his children would surely go to be with their mother. The policeman let him go to the other line.

The line into the lodge went quickly. Tim saw that there was another line moving quickly out of the lodge to the parking lot and buses.

The people in line with him were all in some stage of pain or debilitation from the sun. All exposed skin was reddened, usually blistered. Those without eyecover shielded their eyes from the light. Some needed assistance, apparently unable to see.

It took a long moment for Tim to get used to the relative darkness inside the lodge. They had shades over the windows. He had just started to look around when a hand touched his arm.

He barely recognized Tim, Jr., his face was puffed out and beet red. As he spoke his lips were cracked and bleeding, "Mom's over here with the twins."

Father and son eyed each other. From the look on Timmy's face, Tim Wesley knew the his own face must be dreadful. They turned to push their way through the crowded lodge.

"What in the hell has happened?" Tim asked, not really expecting an answer from his son.

"They have the TV on in the lounge. They say its the Ozone Hole. Something about it expanding up from Antarctica during the weeks of overcast weather. This is the first day of sun and the ultraviolet light is burning people all over New Zealand and they say over in Tasmania and south Australia. too."

The twins were sitting on a couch. Dorothy and an older white haired woman in a white jumpsuit were bending over them. Between the women Tim could see one of the boys, he usually could tell them apart, but not now. The boy's forehead was bleeding and his lower face was swollen all out of proportion. The woman in white was putting some shiny ointment on him. He cried with each touch.

Dorothy felt Tim come up behind her. She turned to him and gasped at the sight of him. The nurse looked up at Tim.

"Your husband?" the woman asked Dorothy.

"Yes."

"Use the rest of this on him," the woman said as she handed Dorothy the small bottle of ointment, before leaving.

"How are the boys?" Tim asked.

"The nurse says we have to get them to a doctor or hospital. And if we can, get to a chemist shop for some more Silvadene cream." She indicated the ointment she was preparing to spread on Tim's cheeks.

"And the girls?" Tim asked. "Ouwww." She had touched his face with the Silvadene.

"I haven't seen them yet."

"I'll keep checking," Tim, Jr. said as he turned away.

"I'll check too," Tim said.

"Just a minute. Let me finish." Dorothy spoke with unusual firmness. With one hand she pushed Tim's face into the light and continued to anoint his blistered face.

When Dorothy finished, Tim turned to see Timmy racing back through the crowd to him.

"Dad, you better come."

"Honey, you stay with the boys," Tim said to his wife.

Timmy led him through the crowd to a vestibule off the entrance. There Tim saw his daughter's bright orange parka, but around her head was nothing but bandages. At second glance he saw Hillary was seated on the floor beside Alice, she was bandaged too.

"You're her father?" a man in ski gear asked. "They're both burnt very bad, worse than most. And they have snow blindness, maybe worse. Are you in a car?"

Tim Wesley nodded without words.

"You need to get them to the Medical College Hospital in Christchurch this afternoon. They'll be busy, but I think they'll take these girls. Tell them Dr. Morris Thurman saw them and referred them. You got that? Morris Thurman referred them."

"Yes. Thank you, Doctor."

Hearing her father's voice Alice cried, "Daddy!"

"I'm here Honey." He lifted her to her feet and carefully hugged here.

"Timmy, help Hillary. Let's go get Mother and the boys."

Tim, Jr. rode up front with his father on the trip back out of the mountains. Dorothy rode in back, ministering to the twins and the girls. Only an occasional whimper or moan broke the silence. Tim tried the radio, but it was only static.

When they left the mountain valley Dorothy took over driving for Tim so he could rest his swollen eyes. Now out of the shadow of the mountains, Dorothy was able to get the Christchurch radio news channel. They learned that the Ozone Hole really was the culprit.

Unbeknownst to the authorities in New Zealand, the thinning of the Antarctic ozone layer, increasingly common over the past decade, had, over a period of three weeks, spread north until its thinnest section encompassed parts of New Zealand and Southern Australia.

Only with the break in the overcast skies after weeks of storm had the Meteorological Service discovered that a good deal of the ultraviolet protection had gone from the atmosphere over New Zealand. And it was, of course, even worse in the high mountains and snow.

Tim Wesley had chosen a poor morning to hit the slopes early.

———

July - Southern Connecticut

After dinner everyone offered to help Molly with the dishes, but she shooed them all into the den again. By the time Andy had finished stoking the fire and the Brady's had settled into the overstuffed loveseat in front of the fire, Molly was back with a tray of four snifters and a decanter of brandy. Andy closed the tempered glass doors on the fireplace and took a seat to the side of the Brady's.

After everyone was served and situated in front of the fire, Andy presented the question he had asked Brady to come to his home for. "Is there any sign that the few changes we have been able to get started have had any effect? The fluorocarbon ban or the emission controls?"

Brady looked at the senator and slowly shook his head. "I wish I could say otherwise, but it just doesn't work that way. The changes we make now won't have full effect for months, years or even decades. But it is a start. Right now we're still getting the full picture of what we have already ruined. The McGill University study on acid rain that the President promised the Canadians we would pay for is just out. It shows that the damage we are doing to temperate and sub-arctic forests with acid rain is worse than what we do with logging operations. And they think the acidification of sub-arctic soils caused by the acid rain from American and Western European industrial plants is increasing the release of methane from the tundra soils. And that appears to be another big factor in the global warming picture."

"Christ, isn't there ever any good news about the environment?" Molly blurted out. The other three looked at her, she shrugged.

Andy smiled and said "Sure, just last week the Malaysian government caught their Minister for the Environment selling indulgences to clear cut teak forests to a Japanese furniture company. It only took four days for them to put him before a firing squad. Unfortunately, they were more upset over the bribes he didn't share than the trees that were lost."

Molly responded, "Hon, I think we need to discuss what we mean by good news."

Andy continued, "Actually, from my point of view, the good news is that we at least have reached a higher level of awareness of the problems. I may still get threatened by automobile union bosses because they are afraid that regulation of the auto industry will cost jobs, but I can take some satisfaction that they are worried we might be successful in stopping the old gas guzzlers."

Before anyone else could speak a shutter on the window in the den banged loudly against the window once and then three more times in quick succession.

"Sounds like the wind picked up a bit," Andy said.

"And if I'm not mistaken, that's the south side of the house." Brady commented, Andy nodded in response. Brady continued, "I think we had better check the TV or radio. As I said, south winds don't bode well when a hurricane is in the neighborhood."

Andy looked at his watch. "Cable news has weather at fifteen after." He went to the small television on the den bookshelf.

He did not need to turn to cable, when the picture came on the local station had a weather warning subtitle scrolling across the screen.

"Damn," Brady muttered as he stood next to Andy and read the scrolling words.

Andy read aloud for the women who had remained by the fire, pausing his reading for the words on the screen to roll by. "Hurricane warning for all areas of Long Island, Metropolitan New York, Connecticut, Rhode Island and coastal Massachusetts. The eye of Hurricane Clark is approximately ninety-five miles south-southeast of Montauk Point, Long Island and headed northwesterly."

When the scrolling started to repeat itself Andy flipped through the channels. Ginger and Molly had now joined them at the television set. Finally a weather program came on with a strained looking weather girl reading from a sheaf of papers.

"All parts of the New Haven viewing area are covered by this storm warning. You are advised to seek adequate shelter and avoid low lying areas. Emergency shelters are being established, check with your local civil defense authorities. If it follows its present track Hurricane Clark is expected to make landfall near Southhampton, Long Island within the next hour.

"If we can have the New England map, I'll show you how this came about." The weather girl paused a moment while the new map appeared behind her. She looked off camera at a monitor and motioned with her hands to the map. "This stationary high pressure ridge running from Canada across Maine and out to the Atlantic apparently interrupted the hurricane's eastward progress away from the coast and this, combined with the low pressure system behind the ridge has drawn the hurricane in to the coast. A spokesman for the National Weather Service Severe Storm Center states that the current path of the storm is likely to pass over Long Island and reach the mainland anywhere from New York City to New London."

Brady was nodding vigorously and muttering something about `they should have known better.' Andy left the television on, but went over to peer out the window into the dark. Able to see nothing, he returned to the others.

Ken spoke, "I guess we'd all better get used to this. Hurricanes like this one and Sandy and especially the really big ones like Katrina, Andrew and Talullah, and the huge typhoons in Asia are really just huge heat engines drawing their power from warm ocean waters. And as the waters keep getting warmer and the warm waters spread farther from the tropics these killer hurricanes will only increase in number and size. And you folks on the east coast are especially vulnerable. That nice warm Gulf Stream out there," he pointed to the east, "is a marvelous runway to launch the Caribbean hurricanes right at you."

Ginger asked, "Isn't there anything your Gaia theory can't threaten."

Ken feigned hurt feelings, "Don't blame Gaia, its not her fault. But no," he polished an imaginary medal with his knuckles, "we haven't found anything so far that our theory can't screw up or imperil." No one laughed.

Molly spoke, "Now what?" She looked at Andy, but Ken Brady spoke first.

"I didn't really notice. How far are we above the water?"

Andy answered, "I'd say a good thirty five, forty feet. Here on top of the hill."

"That should be plenty. I'd say we are as safe here as we would be on the roads trying to get to town. We drove through miles of pretty low country to get here today."

"You're the meteorologist."

"Can you close the shutters? I mean are they real or just decoration?"

"Real, we always close and lock them, at least the downstairs ones, when we leave the place."

"I think we ought to batten down the hatches before the storm hits."

Andy nodded agreement. "You'll probably want to pull your car into the barn, er, garage, too."

Molly pulled two yellow rubber rain slickers from the front closet and the two men circled the big old house by flashlight securing the shutters. Molly and Ginger did the same for the upstairs from inside by leaning out the windows and grabbing the shutters. Everyone gathered together in the foyer when they were done.

"Three windows upstairs have stuck jams. We can't get them open," Molly said as she attempted to get her wind blown hair back in place.

"I'll think we'll be OK." Andy did not sound too sure. "I think I can use that brandy again."

The fire roared behind the glass doors of the fireplace, fanned by the updraft of the high winds blowing across the chimney. The couples sitting in front of it could hear the air sucking into the brass grates below the fireplace doors. The noise of the fire was soon overshadowed by the sounds of the storm.

As the winds rose outside, the conversation turned from the mutual business discussions of the environment by the senator and the scientist to more personal conversation by both the wives and husbands. Relaxed by the brandy and warmed by the feeling of comraderie the couples were soon telling stories on their spouses that they would only tell to good friends, as the others now were.

They checked the television from time to time. As the storm approached all of the local stations went to exclusive coverage of the hurricane. There were reports on the massive traffic jams on the 495 freeway on Long Island and on the Connecticut Turnpike as people fled the path of Clark. The satellite photos showed the track of the ominous spinning spiral of clouds as it did its curious left turn over the Atlantic and headed for Long Island.

The Knowles got a telephone call from their son, Chad. He had seen the reports of the hurricane on television in Texas and wanted to make sure his parents were alright. Andy had taken the call first and then put Molly on the line to talk to her son.

"Yes, dear. Everything's just fine. Your father has the shutters all latched and we couldn't be cozier. We just..... Hello? Hello?..." She looked over to Andy. "Hon. The line just went dead."

Before Andy could speak Brady interrupted. "Yeh, We lost the cable signal, too." He pushed the television button off.

Molly set the phone into its cradle on the desk and no sooner had she done so than the desk lamp dimmed, pulsed and went out.

The four people looked to each other in the firelight in silence. As if on cue, somewhere in the house a shutter vibrated in the wind like the reed of some huge musical instrument.

In a mock eerie voice Andy said, "It was a dark and stormy night...." Molly half smiled.

Andy walked through the foyer and went to the front door. "Let's see what its like outside." He unlatched the door and pulled it open to peer out. Nothing happened for a moment, but then a gust of wind hit and he had to throw himself against the door to keep it from slamming open. Putting his shoulder into the door he managed to shut it. He came back into the den with his face and hair wet and with a sheepish look on his face. He brought the storm lantern they had used for the shutters with him.

"Maybe its time to crawl into a nice warm bed and pull up the covers," Molly suggested, checking her watch,.

"I'm not sure any of us are going to get any sleep, but it is worth a try," Ginger replied. The men nodded agreement. Molly led the way upstairs.

Outside, the winds of Hurricane Clark grew in strength. The two couples lay awake listening to the howling wind, pummeling rain and occasional bang of debris hitting the house. Thunder crashed and lightning flickered through cracks in the shutters.

They had not been in bed for more than twenty minutes when the house gave off a strident creaking sound of timbers and nails under strain. The creaking sound was punctuated by the sound of breaking glass in another room. Andy took the lantern with him. He met Ken in the hallway.

In an extra bedroom they found a tree limb had smashed through one of the unprotected windows. Through the shattered panes rain was coming in. Andy handed the light to Ken and then pushed the branch back out the window. With a bed pillow he pushed the broken glass outside and then reached out through he now open casement and pulled the shutters in and latched them. Soaking wet, he took the light and led Ken back out into the hall.

When they reached the hall, the house groaned again. The two women were waiting for the men.

"It is getting pretty bad." Molly said as she pulled her robe tighter around her.

"I'd say so," was all Andy could respond with.

Another mighty blast of wind sent shudders and creaks throughout the old house. They each felt the house move beneath their feet.

"You have a basement?" Ken asked Andy.

"Not as such, in this old of a house. Just the root cellar that they expanded to hold the coal for the old coal furnace they had before the fuel oil tank went in."

"That'll do. I think we ought to make use of it."

"I think your right!"

Molly opened the linen closet and everyone helped carry extra blankets downstairs through the kitchen. Andy led the way down the stairs shining the light back so the others could see. In the kitchen they picked up another flashlight which Ken took.

The cellar was a mess; old boxes, junk and cobwebs. It was not more than fifteen feet square. Andy pushed a stack of empty boxes aside and the two couples spread blankets and made impromptu beds for themselves. Above them, the wind roared and the house groaned.

"Wow, and honest to goodness adventure," Molly said, and she immediately wished she had not.

"Yeh, `How I Spent My Summer Vacation. By Virginia Brady.'" was Ginger's response as she shivered under her blanket with her husband.

Andy turned the light out, "Just another cozy evening huddled in the root cellar."

"Hey, I've spent twenty years in meteorology and never got to experience a hurricane firsthand. This is really interesting."

"Yeh, right. I've spent twenty years in national politics and never experienced a coup d'etat. There are just some professional accomplishments you can to live without."

They could hear the increasing ferocity of the hurricane; now a constant tumult above them. With every passing moment more debris crashed into the house. Finally, they heard what sounded like dozens of pieces of lumber slamming against the southern wall.

In the dark Andy speculated, "Was that the barn going?"

"Could've been," Ken replied.

The house now let out with a massive groaning creak, followed by breaking glass and cracking wood. Directly above their heads they heard a piece of furniture crash to the floor. This was followed by creaking and cracking that rose to a crescendo.

Andy flipped on the flashlight just in time to see a cascade of dust settling down from the cellar ceiling. The staccato crack of individual pieces of wood breaking merged into a rumble.

As they watched, the stairway walls leading up out of the cellar disappeared in a cloud of dust and falling debris.

Andy instinctively recoiled away from the falling wood and plaster and covered his wife with his body. As the ceiling shifted a beam pulled away from the foundation. It fell and struck Andy a glancing blow on his shoulder. Choking from the dust which now blew around from the wind coming in, Andy struggled to push the beam off of him. It tumbled to the middle of the cellar floor after a solid push from Andy and a pull from Brady.

Andy reached for the light he had lost in the crash. He could see from the fallen flashlight's beam that Brady had gone back over to his wife who was huddled under their blanket, apparently unhurt. Grasping the light, he turned it on Molly who lay beside him.

The beam which had hit Andy in the shoulder had hit Molly squarely on the side of the head. She was unconscious. An ugly gash ran from her temple back into her hairline. Andy did not need to check for a pulse. The slit above her ear pulsed spurts of blood as he watched in the dim light. Ginger, seeing the injury, handed him a cloth which he used to press against the gash.

"How bad is it?" Brady was on his knees beside them. He had to shout to be heard.

"I don't know."

Brady moved the log farther away from the corner were Molly lay. They wrapped her in the blanket and Andy lay next to her to shield her from the wind and rain and anything else that might fall.

The elements now whirled around the cellar through the partially exposed stairwell. Molly's bleeding stopped, but she did not regain consciousness.

The storm continued to roar above them. From time to time they heard part of the house rubble scrap by above them and then blow away. Debris continued to roll down the stairwell.

With nothing to do but wait, the foursome lay in the cellar until the storm died out.

———

The hurricane took as long to die out as it had to build. The worst of the winds lessened first, but the torrential rain continued unabated for the rest of the night. A large puddle of water formed at the base of the stairs.

Dawn came slowly. The steely gray light that pierced the last of the storm clouds made its way into the cellar. Ken Brady went up the litter-strewn stairway and stuck his head out of the largest hole in the debris covering the cellar entrance.

Andy looked up to him and asked, "How is it?"

"Damn," was all Brady would say.

Andy looked down at his still unconscious wife. Ginger was now next to him.

Ginger spoke, "Go on up and check it out, I'll stay with her."

After a slight hesitation, Andy followed Ken up the stairs. Ken was pushing at some boards that blocked the top of the stairs. They shifted out of the way and Ken squeezed up through them into the morning air. Andy followed.

Andy had tried to prepare himself for what he might see, but even though contemplated, the shock of the devastation was nevertheless there. As he pulled himself up through the hole in the debris he could see the rubble of both the barn and house tossed as far as he could see. Trees were broken and twisted. Both his mini-van and Brady's rental car were smashed under the collapsed barn.

To the north and west the contents of the house were thrown along with an occasional piece of furniture. Most of the house was simply folded sideways and collapsed upon itself to the north. If the entrance to the cellar had been on the other side of the kitchen they would not have been able to get out. The roof had blown clear of the rest of the house and rested fifty feet away, still intact, but upside down. The wedge of the roof looking like some crazy boat grounded on a shore.

Looking down the hill toward the river, Andy could see neither boathouse nor boat. The verdant woodlands beside the river, which he had admired two days before, now stood haggard and bedraggled. Wind whipped tree limbs lay everywhere. As he looked about him the sun came out of the now docile clouds in the east.

Ken had gone to check the cars. Andy started to walk over to the barn, but Ken started back before he got halfway there.

"The main roof beam fell right straight down and got both cars. Even if we could get the crap off, they're probably too smashed to use." was Brady's report.

"I can't see the boat, So I guess we are stranded 'til the cavalry comes. Let's get Molly and Ginger up into the sun."

The two men walked back to the wreckage of the house.

"When we get Molly out what are we going to do with her."

Andy thought and then moved off, circling the wreckage. Seeing what he wanted he pushed aside lumber and started pulling on something. Brady went to his side and helped him pull one of the mattresses out of the rubble.

With the mattress, they returned to the cellar entrance. They had cleared a hole big enough to carry Molly out when they heard the thumping whir of helicopter blades in the distance.

Governor Morgan Drexel felt out of place.

The National Guard jumpsuit did not fit the aging politician very well. And, the governor avoided helicopters whenever possible. But the disaster provided a wealth of "media opportunities" as his campaign press secretary had put it. The decision for the helicopter tour had been insured when the cable television news crew had asked if they might accompany the governor on the chopper ride through the worst of Hurricane Clark's devastation. Not to be upstaged, one of the local network affiliates decided to come along in their own helicopter. The media coverage of a disaster like this was worth its weight in gold to a politician up for re-election. Governor Drexel would be foolish to miss this opportunity.

The plan was to follow the coast up to New London, refuel and have a press conference and then turn back west and follow the coast down to Bridgeport.

They were almost to East Lyme when the governor thought of his hosts at lunch the day before and asked the pilot to take a side trip.

————

Governor Drexel's press secretary was in ecstasy. You just could not manufacture press coverage this good. Hurricane Clark had devastated the entire New England seaboard, just as Sandy had done years earlier. Hundreds and thousands of summer vacationers and residents had been caught unaware on Martha's Vineyard and eastern Long Island. And immediately after Hurricane Clark hit New England, Hurricane Deborah followed the track Hurricane Andrew had a few years before and walloped southern Florida. In all the horrible news there has only one bright spot, a human interest story with Governor Drexel as the hero.

Every news outlet in the country had coverage of the governor's helicopter landing in the ruins of the nationally prominent senator's demolished country home, rescuing the senator's comatose wife and bringing the other survivors to safety. Interspersed with other footage of the utter destruction of coastal Connecticut, New York and Florida, the clips of the governor's good deed were a press agent's wildest dream. He had even been able to get television coverage of Mrs. Drexel, wearing a Red Cross volunteer's jacket, going to visit the recuperating Mrs. Knowles in the New Haven Trauma Center.

————

August - Garo Hills, Bangladesh

The sound was unmistakable. It echoed through the verdant valleys, an intrusion in the pastoral life of the village. Like any military man, Akim Darpheel had no trouble recognizing the sound of an armored column on the move. It was not unexpected.

After the turmoil of the spring, life had settled into a calm routine over the summer. Contrary to his fears, he and his comrades from Dhaka had been welcomed back to his families village in the Garo Hills, but for a reason he had not anticipated.

Like the rest of Bangladesh and the neighboring countries, the Garo Hills and the lower Assam region on both sides of the border were beset by hordes of refugees from the lowlands of Bangladesh. They had started coming into the hills looking for food and shelter before Akim and his entourage had arrived. At first Akim, Latifa, her children and the other two soldiers had been viewed as much the same as the common refugees, even by the relatives of Akim. However, in the first week after their arrival in the village, Sergeant Quist had confronted a party of thieves trying to steal a calf from Akim's father in the middle of the night. Quist and Lieutenant Mullahwi had fired their rifles and scared the thieves off.

It turned out that every farmer in the village and surrounding hills was having trouble with starving bands of refugees breaking into their homes and storage sheds at night. No one had seen anyone from the police in Jamalpur or Mymensingh since the storm. Word spread of Quist's confrontation with the thieves. Soon the village mu'azzin, who also served as de facto mayor of the village, conferred with Mahmal Darpheel and the other elders and together they offered Akim, Mullahwi and Quist the opportunity to work as guards or police for the village and the farms of the valley. Akim and Latifa were given use of a house in the village and Quist and Mullahwi were put up in the Mu'azzin's home. When the farmers took their crops to market in Jamalpur they took two of the soldiers with them for protection.

So, life had proceeded throughout the summer. News from the outside world was scarce. They sometimes brought newspapers back from Jamalpur when they went to market, but more often than not, there were no newspapers. For all intents and purposes the government and much of the commercial life of Bangladesh had ceased to function. When they could find batteries the villagers often listened to the BBC, Voice of America or one of the Indian radio stations; the stations in Dhaka had never gone back on the air after the typhoon.

After the move on Chittagong, the Indians had not taken any action for a while, but the stream of refugees became unbearable. The Indian Army had first set up a defensive line in the delta to prevent to millions of refugees from the delta from pouring into the Calcutta metropolitan area, which had problems enough of its own. The frontier of the two countries was only twenty-five miles from downtown Calcutta, one of the world's most populous and certainly its poorest metropolis.

Then, in July, word had spread that the Indians were asking the United Nations Security Council to act in stemming the flow of refugees, threatening to take action unilaterally if the U.N. did not act. The Islamic nations in the U.N. had blocked any move by the U.N. interfering with the sovereignty of a fellow Islamic nation, Bangladesh. Apparently the Bangladeshi ambassador was maintaining the facade that a nation state still existed in Bangladesh.

Finally on the last trip to Jamalpur with a tea shipment Mullahwi had heard that the Indians had occupied Jessore, Rajshahi and other border cities and that the Indian Parliament was considering a resolution annulling the 1947 British partition of Bengal between Hindu and Moslem sectors. If unchallenged, this would have the effect of annexing Bangladesh to India. Akim, Quist and Mullahwi knew from personal experience that there was not enough left of the Bangladesh Army to challenge anything.

Now, with the sound of tank treads and growling diesels echoing down from the uplands near the Indian border, it was clear to Akim that the Indians were making good on their threat. The old colonial road from Tura on the Indian side to Jamalpur and Mymensingh passed through the valley. It was one of the few roads through the Garo Hills. The Indians would want to come down to Mymensingh, with its road and rail lines to Dhaka through the line of least possible resistance and population. That happened to be the old road through Akim's village in the Garo Hills. In fact, his village would be the first settlement inside Bangladesh which the column would pass.

Not knowing exactly why, Akim went to the house and put a clean uniform jacket on. Latifa had brought her dead husband's bags along with her and the extra uniforms had come in handy. Akim, as well as Quist and Mullahwi, had gotten into the habit of wearing their uniforms in their police and guard duties. It seemed appropriate.

As Akim walked from the village out to where the road curved out of the hills, he saw the rest of the villagers were also going to see the Indians come through. Quist and Mullahwi were already at the road, also in uniform.

When Akim came up next to him, Mullahwi asked, "What should we do?"

Akim thought, smiled and said, "Well brother, you can either smile at them and wave. Or you can charge the first tank with your bayonet fixed and see if you can scare them back across the border."

Quist laughed. Then Mullahwi and Akim joined him.

The Indian column could now be seen coming around the upper bend of the road. The column was led by two British Chieftain tanks and several armored scout cars of Indian make.

The Indians continued until they were at the first bend above the village. There the tanks stopped. Two scout cars pulled around the tanks and approached the village.

On either side of Akim there were perhaps two hundred villagers. They stood silent as the Indians approached.

One scout car stopped about thirty meters out. The other continued until it was only a ten meters away, directly in front of Akim. Akim could see the goggled officer inside the front portholes.

Akim heard the chirp of a loudspeaker being turned on. From public address speaker on the scout car, a voice, in near perfect Bengali, boomed out at them, "Lay down your arms!"

Akim looked to his comrades, then he dropped his pistol belt to the ground and stepped back. The other two did likewise with their rifles.

"Have your commander approach our vehicle, alone!" was the next command.

"That's you, friend," Mullahwi said to Akim.

"Why me."

"But this is your home village and you are wearing the major's uniform."

Akim turned his head and looked down at his shoulder. He was indeed in a major's uniform. Latifa had promoted him to her dead husband's rank when she washed the uniform. After a moment's hesitation, Akim strode forward.

Akim could see that the machine gun port was trained on him following his progress toward the scout car. When he was a few paces from the vehicle the hatch popped open and the officer poked his head out. "Do you surrender your garrison?" the officer asked in perfect, albeit Hindi-accented Bengali.

Akim Darpheel could not help but laugh. "Certainly, I surrender my entire garrison, all three of us."

"There are only three of you."

"That is right and we are all the military personnel you will find in the next twenty kilometers, if at all."

The officer seemed disappointed. He apparently was intent on the glorious conquest of the Bangladesh nation. Akim remembered the dreams of such martial action. The young Indian officer was in the moment of military glory and he found himself faced by three scruffy would-be enemies and a couple hundred curious villagers, instead of the stalwart foe his life's work had trained him to defeat.

After a short hesitation the Indian said, "Well, we will be occupying your town and setting up defensive positions south of here. Please assemble the residents in the center of the village for an announcement by our civil affairs officer. Understood?"

"Understood."

"And one of my men will collect your weapons."

"Is that necessary?'

"For the time being, of course."

Akim nodded and turned back to the village. Behind him he heard the crackle of radio static as the Indian officer reported back to the column.

Before rejoining Mullahwi and Quist at the side of the road, Darpheel went over to where his father and the Mu'azzin were waiting, watching the events transpire on the road. Darpheel told them of the orders of the Indians. The old men nodded and said they would gather the people.

By the time he got back to the other two, everyone had retreated from the edge of the road to avoid the dust of the passing of the tanks and trucks of the convoy. The bulk of the convoy took several minutes to pass.

"I count at least three brigade placards. This is a full division." Quist said.

"Probably the Third Armored, out of Shillong. They came into Sylhet in '71 to support Rahman's independence declaration. We often thought that their real purpose in Shillong was to move against northern Bangladesh when the time came." Mullahwi had worked in "Enemy Order of Battle" in the Intelligence section of the general staff.

"It seems the time has come." Darpheel said.

Toward the end of the convoy there were several staff cars in a row. As they watched, one of the staff cars, after it passed them, skidded to a stop beside the road. When the dust had cleared, the door to the car opened and a man got out. He was an Indian general officer. There was something familiar about him. He walked toward them.

It was Quist who finally recognized the man and snapped to attention in a salute. A second later Darpheel also recognized him, his crisp uniform and grooming had improved since they had last met.

"Colonel Majlis!" Darpheel said as he, too, saluted.

Majlis smiled and returned the salute. "Ahh, but I am no longer just a colonel. But then, you are no longer a just a Captain, eh, Darpheel."

"Latifa gave me Radija's uniforms."

"Yes, how is Latifa?"

"She is fine. We have a house here."

"Perhaps I could pay her a visit."

"Certainly, Colonel. But what is this uniform? The Indian Army?"

"I went north to Saidpur after we parted. I was there when the Indians occupied the rail line and city. They declared me a prisoner of war and sent me to a camp in Punjab. I was the senior Bangledeshi officer in Indian hands and when they decided to take control of all of Bangladesh last month they consulted me. One thing led to another and they decided that it would be best if a Bangladeshi was the token leader of the occupying forces, a puppet government, at least for a while. Last week, I found myself in Delhi being sworn in by the Indian President as the Governor of the Indian State of East Bengal."

"Bismallah!" was Mullahwi's startled response. Akim and Quist simply stared at their former commander with open mouths.

"Come, let me see Latifa. There is much to talk about." Majlis signaled the staff car to follow him as he walked into the village with the three incredulous Bangledeshi soldiers. No, they were now simply Bengalis, East Bengalis. And Majlis's problem of finding trustworthy men to assist him in his puppet government had become easier.

———

August - Boulder, Colorado

About twenty miles south of the National Center for Atmospheric Research and GAIA's headquarters in Boulder, Colorado is another government agency with a very similar mission to NCAR and GAIA. The U.S. Geologic Survey (USGS) has its national seismological information center in the city of Golden, near the Colorado School of Mines. Like NCAR and GAIA, the USGS office is a clearinghouse and focal point for data from all over the world. Whenever an earthquake takes place anywhere in the world, this office is looked to as a source to provide the epicenter location and relative strength on the Richter scale for the quake.

The work of the USGS office in Golden did not relate directly to the work of GAIA, except in two areas. The scientists of GAIA had a standing request in for any information coming into Golden on any volcanic activity anywhere and any seismic activity that would affect oceans or icepacks. The volcanic activity could greatly affect the level of greenhouse gases in the atmosphere and the amount of water vapor in the air, which also has a greenhouse effect. It was the latter request that led to the telephone call Ken Brady received in mid-July.

"Dr. Brady?"

"Yes."

"This is Claire Bergen with the USGS in Golden."

"Yes, what can I do for you?"

"We have a request from your office for any information on any significant seismic or volcanic activity in Antarctica."

"Yes?"

"Well, we have been getting reports for some time on certain low level activity in the Antarctic area. We haven't brought it to your attention since it appeared to be subcrustal magma or anomalies. There are several dormant volcanoes in Antarctica, including Mt. Erebus, which was very near the readings. We didn't consider these to be significant."

"That's probably right, unless there is actual volcanic activity or seismic disturbance it would not have any effect on our work." Andy was wondering what she was getting to.

"That is what we thought, too. The readings came from the British and Australian Antarctic missions and until recently their seismic equipment was rather limited. However, some months ago the British installed some newer broad spectrum seismic gear and we have just gotten around to analyzing what it is giving us."

Claire Bergen continued, her voice somber and overly professional, "It appears that what we thought was low level activity, was really fairly active, but at a frequency that did not pick up well on the old seismic gear. Now both our analysis and that of the British indicate that it is not normal seismic activity and not magma either.

"We have plotted the activity, frequency and timing on a map overlay of the area. We are sending it up to you by courier right now. Our best guess is that the high-frequency seismic activity is on a rock/ice interface in an area that roughly approximates the Ross Ice Shelf."

"My God!" Brady cut in.

Bergen continued, "We think the Ross Ice Shelf or at least a big part of it is sliding north."

"By sliding, I assume you mean something more than the normal glacial movement."

"Of course, glacial creep is of a totally different frequency and magnitude than what we are talking about. What the Brits have found is a wholesale slippage of massive formations of ice on a scale that rivals a tectonic plate shift."

Ken Brady concluded the call and hung up the phone in a state of shock. Claire Bergen's prognosis was the worst nightmare the GAIA computer could conjure up.

The Ross Ice Shelf was a glacial ice pack roughly the size of France that was grounded on the coast and continental shelf of western Antarctica. Unlike the Arctic and Antarctic sea ice that floated on open ocean and displaced almost exactly its own volume of ocean water, the Ross Shelf was, for the most part, setting atop the solid rock below it. In places the Ross Shelf measured in the thousands of feet, being an extension of the massive continental ice packs of Antarctica.

If the Ross Ice Shelf slid from its resting place on the coast of Antarctica it could raise sea levels by twenty to forty feet around the world, instantly. Even a slow slide would cause greater melting of the underlying ice from friction with the rock and the melting ice would lubricate the slide. And as the oceans rose slightly, as they already were from thermal expansion even without the Ross Ice adding to it, the part of the ice shelf resting on the ocean floor would float higher and precipitate an easier and faster slide for the landlocked part of the shelf.

Claire Bergen had only mentioned the Ross Ice Shelf. There was another huge ice shelf, the Ronne, on the other side of the western Antarctic region and several minor shelves.

The breakup or drifting of the West Antarctic ice shelves had been postulated by some fairly eminent scientists as early as 1968, but such an occurrence was of such colossal magnitude that it was not considered by most scientists to be of concern in the foreseeable future, barring the unforeseen.

Any way you looked at it any mention of the Ross Ice Shelf sliding was a true nightmare of global proportions. Ken Brady thought for a minute and then picked up then phone. Washington had to know about this immediately. Then, Ken brady put the phone down, this was so staggering in its importance that he would have to wait and see the USGS charts. He did, however, assemble the senior staff to tell them and call his friend, John Colton, fellow GAIA researcher with the responsibility for ice pack and ocean information in California.

———

August - San Diego, California

It had been an idyllic morning for John Colton. At least the part before he had to get into the car and head south on Interstate 5 to La Jolla and the Institute.

He had been awakened by the sun beaming in the open french doors from the bedroom out to the balcony. The air was fresh from the heavy rains the day before. Their canyon edge home in Solana Beach was situated so that morning sun came in the bedroom and evening sunset over the Pacific could be seen from the family room balcony. Colton's wife, Becky, had used her position on the sales staff of the developer to choose the best lot and house choice in the development.

After awakening, John Colton had rolled over and was greeted by the sight of Becky, only half covered by the bed sheet, sleeping peacefully beside him. Neither he nor Becky usually wore nightclothes, so her beauty and physique was arrayed before him in the morning sun. Only the long brown hair interrupted the white on white of skin and linen sheets. With his big toe, John reached over and hooked the edge of the sheet, pushing it lower, exposing the most fascinating part of her body, at least the currently most fascinating part.

The sheet slid easily off of Becky's high rounded belly. He watched intently for a long moment to see if he could actually see the movement Becky had let him feel in the belly. They had learned only the week before through amniocentesis that the child would be a girl. He could see no movement, perhaps his daughter was sleeping along with her mother.

When he could resist the sight no longer, John Colton reached over and gathered Becky into his arms, turning her so that her backside molded into his body and so he could reach down and feel the rounded belly. His efforts were rewarded with a gentle kick, or was it a shove? John then slipped back into that very rewarding sleep that only comes after once awakening on a sleepy morning. It would be a herculean task to respond to the alarm and get out of bed.

The ride from Solana Beach down to the Scripps Institute in La Jolla was really not that far to drive each morning. The only problem was the volume of traffic. Even leaving before seven o'clock did not help some mornings, and this was one. The six southbound lanes into San Diego had been bumper to bumper for the whole ten miles until he exited. And then the morning crowd into the University and on further to the Institute on the coast above La Jolla was gruesome. At last he turned off into Scripps and parked.

John Colton had dreamed of being an oceanographer since he was a child. And if one dreamed of being an oceanographer, the place one dreamed of being was Scripps Institute. Colton rarely thought of this without wondering at his good fortune to actually have fulfilled such a dream.

There are a handful of institutions whose very name defined their stature in their academic discipline. What the names Mayo, Wharton, Juilliard, West Point, and M.I.T. were in their fields, so too, was Scripps Institute in the field of oceanography. Colton was proud to be a part of Scripps. He realized how rare it was for a man to go to work, knowing that he would enjoy every minute and be intrigued by every facet of his day's work. This was even more the case in recent months, when it became clear to all who worked on the GAIA project that they were key players in a global drama.

Colton had spent the morning going over reports on the latest series of temperature readings from the oceans of the world, his specialty. Everything confirmed the El Nino effect had spread fully half a globe across the Pacific from the hottest spot off of Indonesia to the waters of South America. His staff had already fed the data into the computer system and Colton had the first opportunity to see the results. The fact that the President had now assigned U.S. naval vessels and military aircraft around the world to assist in gathering the data that Scripps had formerly been gleaning from haphazard samplings from a handful of research ships and a few satellites was a significant advantage. Unfortunately, it appeared that the expanded data collection was only confirming GAIA's earlier forecasts, continued and increased heating of oceans and air masses. Parts of the Pacific and Indian Oceans appeared to be up over nine degrees from the expected norms for summer months. Some, hopefully anomalous, readings showed twelve degrees. Knowing that an increase in temperature alone, even without any ice melt, could raise the levels of the oceans by the simple process of thermal expansion, Colton and his staff were hard at work evaluating what the new ocean heating data meant for ocean levels and integrating the British and Russian theories into their analysis. Simultaneously they sent the temperature data to Boulder, for use in the weather prognostications.

Then, just before noon, Colton had gotten Ken Brady's call relaying the news from the seismologist about the Ross Ice Shelf.

"My God!" Colton had exclaimed.

"That's just what I said." Brady replied. "The charts from Golden have just arrived here. Midge and Ted are looking it over. Bart Jastre was home with the flu, but he's coming in. John?"

"Yes."

"What is our latest info on what effect the Ross slide would have?"

"Actually, we haven't found anything to discount the old estimates of Hollin and Barry from '79. The Ross Ice Shelf has a sea level equivalence of about 5 meters and the Ronne and the lesser shelves have about another three and a half to four meters."

"If they let go we have an instant fifteen to thirty foot sea-level rise."

"That's about it. But it might be that only one is unstable. But the chances of that are pretty slim. Even if the Ronne shelf were more stable, the floating force of the extra fifteen feet of sea water caused by the Ross slip on the Ronne would probably float the Ronne also, or at least loosen it up."

"What exactly do we have to confirm or refute the seismic data?"

"Plenty to confirm it. Huge icebergs. Ice pack temperature inside the Ross Shelf last year was barely below freezing and warmer the deeper you go. Not much at all to refute."

Brady paused for a moment and continued, "Well, I'll scan you what I can of the Ross information and express mail the rest. Can you work up a plan of action to get on-site confirmation of any movement and a better look at the temperatures?"

"Can do."

"We'll talk later this afternoon when I know more."

After the call it took Colton a moment to digest the import of the call. He immediately went to find the Antarctic charts and the sea level data he had spoken to Brady of. As he reread the data for the third time he caught himself thinking about the altitudes of places he was familiar with.

His house was obviously safe on the bluff in Solana Beach and Scripps was probably OK on the hillside in La Jolla. Downtown San Diego and the lower buildings below the main Scripps facility were definitely in for it if the Ross gave way.

As he was thinking of this, the next unsettling thought came to him. The problem was not merely the rising water. If the Ross gave way it might do it with a surge that would cause a tidal wave or tsunami that would make all previously recorded waves look tame. Not so much an ocean wave as it was a shock wave in the water, a tsunami could travel at four or five hundred miles an hour in open ocean and when it hit land the shock wave could convert into a mountain of rolling water that inundated coastlines for miles inland. Usually associated with earthquakes, Colton assumed that the tsunami effect would be the same for the Ross Shelf as big as it was only perhaps longer in duration since the sliding of the huge shelf would certainly last longer than an earthquake. Colton made a mental note to refresh his memory on tsunamis and put the staff to work on the potentialities.

———

August - Washington, D.C.

Unlike the rest of the White House that was under the protection of the Secret Service and the civilian uniformed guard force, you had to pass through a Marine sentry to enter this sub-basement of the West Wing. A relic of the Cold War, the Situation Room was the domain of Admiral Kirksey and the President's National Security Staff. As the Gaia Crisis (which the environmental situation was now called within the government) progressed, Kirksey had joined with Dr. Fairchild in arraying the Situation Room with the information necessary to brief the President, his revamped Council on Environmental Quality and the Cabinet on the environmental events and situations. Where once charts of nuclear strike forces hung, there were now maps of rain forest clearing and global isothermic projections.

Andy Knowles could not help it. He could not meet one of these Marines in the White House without thinking of Ollie North. The crisp khaki and blue uniforms and bristly haircuts reminded him of the most famous Marine ex-employee of the White House. Perhaps it was the uncompromising sternness of the sentries that made him think of this each time they demanded he produce the laminated plastic White House photo ID he had been issued when the President had appointed him to the Council. Unlike the civilian guards upstairs who let him pass on his own recognizance, Kirksey's Marines always required identification from everyone except the President. Even Kirksey kept the ID around his neck when he was in the Situation Room. Having gone to Yale Law School instead of into the military, Andy always wondered about the level of suspicion and wariness that seemed to permeate military thinking and procedure. However, as he grew older and spent more time in Washington, Andy had come to learn that a little suspicion and mistrust was a good thing.

The call from the White House switchboard had come in before Andy was out of bed that morning. Molly had taken the call and hurriedly passed the phone under the covers to Andy when the operator identified herself. There had been several staff briefings since Andy had joined the Council including a one in which Andy was called on to brief the Council on the status of the legislative package, but since the first meetings at the White House in March, nothing had been called on an emergency basis.

The Situation Briefing Room was laid out in three tiers of seats, auditorium style. The lowest tier was in front with wide tables with phones and computer terminals. This tier had black leather swivel chairs. A Presidential seal was embossed into the back of the chair in the middle. The upper two tiers of seating were more spartan, but comfortable. They each had a phone set and computer screen with keyboard inset in the desktop in front of them. Andy took a seat in the third row before looking around the room.

He was surprised to see his friend, Ken Brady, here from Colorado and talking with Kirksey and Fairchild on the far side of the room. Behind him stood one of Brady's people from Boulder; the oceanographer, Midge Carter.

Andy recognized several members of the Council from previous meetings, but there were many others he had not seen before. A cluster of Navy officers gathered around what looked to be a map of the Antarctic. Another group of new faces surrounded a man Andy recognized from a Senate confirmation hearing, but Andy could not place the face with a name or office.

The growing murmur in the room died suddenly when the Secretaries of Defense, State, Interior and Transportation entered the room and went to the front row of seats. This was not going to be just a meeting of the advisory council. The HUD Secretary followed closely behind them with his crutches. Andy had heard that the Housing and Urban Development chief was nursing a ski injury.

Andy thought for a moment of going down and speaking to Brady before things got underway. However, he saw that the cabinet members were not seating themselves in the chairs they had gone to. That could only mean one thing.

Kirksey walked quickly to the outer door and announced, "Ladies and Gentlemen, the President." Everyone stood. The President did not attend such advisory body meetings; it was up to staff to brief him on the goings on of bodies like the Council on Environmental Quality. Something big was afoot.

The President was followed by the Vice President and the Chief of Staff. They went to the center of the front row. The three people added to the middle of the row caused everyone to move down. The odd man out at the far end of the first row, the Secretary of Interior, got bumped to the second row. Such is politics, Andy thought.

With the arrival of the President everyone who was not already there went to their seats. Doctor Fairchild went to the podium in the front of the room. Fairchild waited for everyone to settle down and began speaking.

"Mr. President, ladies and gentlemen. There has been an extremely significant development and the President has requested that all of you here today be briefed fully so that you can give him your recommendations for action.

"The U.S. Geological Survey Seismic Office in Colorado has reported significant seismic disturbances in the area of the Ross Sea in Antarctica. Although originally thought to be geologic anomalies, after subsequent analysis they appear to represent previously unexpected movement of the Ross Ice Shelf. This movement could have disastrous effects on ocean levels and climate worldwide. Dr. Kenneth Brady of the GAIA Office and his assistant Dr. Margaret Carter are prepared to brief us on the situation."

As Brady made his way to the podium an unseen motor whirred and the curtains behind the podium parted to reveal a large projection screen. Momentarily a huge map of Antarctica appeared. Midge went to a computer terminal to the left of the screen where she bent down over the computer and as she moved a computer mouse there, an arrow on the map projection moved.

Brady spoke. "The Ross Ice Shelf," he paused for Midge to circle the pointer over the map area, "is a massive block of ice grounded on the continental shelf of Antarctica. It is one of several ice shelves on the fringes of Antarctica, the most significant of which, other than the Ross Shelf, is the Ronne." Midge marked the Ronne Shelf.

"These ice shelves differ from other polar ice in that they are neither sea ice floating in the ocean nor glaciers grounded on solid earth. They are located in areas which would, but for the presence of the ice, be open ocean, however the ice shelves are grounded on the ocean bottom and extend up thousands of feet in some cases above where the sea level would be.

"If sea ice, such as icebergs or the ice at the North Pole, melts it does not affect the level of the water body it floats in. Its floating mass exactly displaces the volume of its melted water. Anyone who has ever watched a stack of icecubes melt in a iced tea glass knows this to be true.

"However, when an ice shelf or glacier melts or slides into the ocean, since it is sitting on the ocean floor and the sea level is nowhere near where it would be if the ice shelf were floating, the ice shelf displaces an enormous quantity of ocean. To be exact, six sevenths of the height, or rather the mass, of the ice shelf above sea level. The surface elevation of the Ross Ice Shelf is in places between seven and eight thousand feet above sea level." Brady paused for the murmur of conversation to die out.

"One of our greatest concerns in this environmental crisis is the issue of global warming. A lot has been speculated about the subject and this topic is one of the prime factors that the GAIA system was meant to study.

"We have heard of global warming estimates of anywhere from no change to six or seven degrees on the average worldwide. For purposes of my discussion today the total amount of increase isn't important. What is important for you to understand is that all of the figures bandied about are averages. And average implies that various parts are different from the norm. That is, all areas do not have identical increases.

"In the case of global temperature we know from experience that some temperatures are stable and hard to change and some are quite variable. A desert may go from scorching in daytime to frosty at night, while a body of water stays relatively even for long periods of time. We also know from experience that when we add heat to a closed system that the coldest parts warm up more than the hottest parts. When you put ice cubes in a pan of warm water and put it on a stove, the icecubes melt before the warm water starts to rise in temperature and boil. The average temperature of the contents of the pan is increasing as the heat is applied but the maximum temperature stays the same until the cold parts catch up.

"What I am getting at is that the earth is somewhat of a closed system, a bit more dynamic than the water in the pan, but it reacts much the same. We know for a fact that our global temperature on the average is going up. And when the global pot is stirred by our climate it is only a natural law of physics that causes the coldest parts to have the greatest increase in temperature. This is exactly what we see happening.

"Over the past several years the temperatures in both the arctic and antarctic have risen much more than the average. The seismic activity Dr. Fairchild spoke of reached its highest point at the end of last year's summer period in Antarctica in April. We believe that there is a good likelihood that the increase in ambient temperature has been just enough to trigger the increased movement of the ice shelf. If you have seven thousand feet of anything sitting in a pile you will have the bottom of the stack warmed by the mere pressure of the weight above it. Witness the increased temperature as you go deeper into the rock of the earth or the lessening of temperature as we go higher in the atmosphere. In the case of the ice shelf we have known for some time that the bottom layers of the ice were only slightly below freezing despite the sub-zero temperatures in Antarctica for much of the year. That was taken for granted. Our problem is that if you increase the ambient temperature of the top of the seven thousand feet of ice, it is only a matter of time before that temperature gradient shifts to the lower levels.

"The ice shelf, like any glacier, exhibits properties of both solids and liquids. We are afraid that the temperature increase has been just enough for the liquid glacial flowing effect to be increased. When the glacial effect starts the massive ice shelf moving, the friction of the ice on the ocean floor melts the bottom layer and this lubricates the ice shelf movement. Then the inertia of the massive movement takes over and the inherent solidity of a huge chunk of ice causes the entire ice formation to lurch forward. It appears that this lurching is what the USGS has reported. In effect you have a block of ice nearly the size of Texas," Midge pushed a button and the outline of the Ross shelf on the screen was superimposed over a map of Texas, "sliding along the bottom of the ocean on a cushion of melting slush. On top of this, because of the location of the ice shelf we have the leading edge of the ice shelf warmed and lubricated by warm ocean waters.

"Dr. Carter, our oceanographer, will now explain the potential effects of this phenomenon."

Midge adjusted a microphone at her computer screen. "Although there are a great number of variables involved in this, we believe that we have a good idea of what is happening, the real questions are when and how much.

"After plotting the seismic information we have correlated it to the available LandSat photos of Antarctica. We have found a rough match up of the areas of movement with pressure ridges in the ice shelf." She pushed a key on the computer and an image of the Ross Ice Shelf from space was shown with longitudinal ripples highlighted and small red bulls-eyes marking the seismic events.

"We are unable to tell exactly, but it looks like an area constituting approximately eighty-three percent of the shelf is moving. The wedged shaped area behind Ross Island and some other peripheral areas do not show any sign of movement, but if four-fifths of the shelf moves there is a strong possibility the rest won't be far behind.

"Based on a study from 1979 by Hollin and Barry, the Ross Ice Shelf has a sea level equivalence of about 5 meters and the Ronne and the lesser shelves have about another three and a half to four meters. Therefore, if the part of the Ross Ice Shelf which is moving were to slide into the ocean it would cause an immediate rise in sea level world-wide of perhaps twelve or thirteen feet with a likelihood that at least some of the other shelves could let go when the Ross does. This is because the other shelves will probably have much the same temperature characteristics as the Ross and the additional floating force of the Ross raising the water will exert great force on the other shelves to lift themselves up. Additionally, if the Antarctic ice shelves break free, the rising oceans could cause similar effects in the lesser shelves of Greenland in the northern hemisphere. The bottom line of this is that if the ice shelves are lost to the oceans we could face a increase of sea level of up to thirty-five feet within a few years, if not immediately.

"There is a lot more we need to know about the ice shelves and that will be a part of our recommendations. I understand that some other agencies are prepared to analyze the effects of the increased sea level. The last thing we need to talk about is the timing.

"The breakup of the Antarctic ice shelves has been a fear ever since we first started to consider the possibility of global warming, but we always talked of the inception date in terms of decades or, more usually, centuries. However, we also used to speak of the time when our industrial enterprises would create entire storm systems of acid rain as a nightmare that was decades away and we now know they occur every day throughout the world.

"The fact is that we don't know enough to give a timetable for this problem. We do know that the movement occurs primarily in summer, November to April. There is a possibility that it could occur in the next warm period this year. It might take years. We might be able to reverse or stabilize the warming to halt the slippage. We do know that if the increase in average global temperature continues as it does now, the Ross Shelf will continue to slide and it will eventually flood our coastlines up to thirty plus feet in elevation worldwide. The important thing is that a catastrophe of global proportions is occurring in slow motion under the Ross Shelf and we may be able to stop it."

From the main podium, Brady nodded to one of Kirksey's staffers, a young woman Marine, who took a stack of papers and started handing them out. Brady waited a moment for the papers to be distributed and then began again.

"These are maps, prepared by a joint team from the Defense Mapping Agency, USGS and several other agencies, which show various areas of our country and the world with both five and ten meter inundations, that's sixteen and thirty-two feet. You will all please note that in accordance with the President's directive after the Miami riots that we protect matters such as this until its release is authorized, the classification on these charts is Secret. This is not so much because the information is confidential, but because the disclosure of our concern for it could backfire on us as it did in Miami.

"Dr. Fairchild...."

Fairchild introduced the next briefers, the C.I.A., who would handle the international situation if coastal inundation occurred. Andy noted that the official he had not been able to recognize earlier was a CIA Deputy Director for Research and Analysis.

While the CIA staff went over lists of foreign cities which would be submerged by five and ten meter increases, Andy thumbed through the maps, They were much more well defined than the computer charts he had gotten in Boulder last March. Obviously a lot of inter-agency effort had gone into the preparation for the eventuality of the coastal inundation.

The CIA briefer was speaking, "There is a strong likelihood that aside from the inundation of coastal cities many nations will experience the almost total breakdown of governmental control, similar to what occurred over the last six months in Bangladesh, when the inundations occur. Of course, island nations such as the Maldives and many Polynesian countries will simply cease to exist. Some entire areas could be in chaos. For instance, in Southeast Asia, all six major cities, Rangoon, Singapore, Bangkok, Phnom Penh, Saigon/Ho Chi Minh City and Hanoi will all be submerged, destroying the entire governmental infrastructure of the region."

One of the Navy Admirals interrupted, "Wait a minute, I've read reports on bombing runs into Hanoi in the Viet Nam War. It's got to be seventy miles inland from Haiphong. And Phnom Penh is at least two hundred miles up the Mekong River."

"That's correct Admiral, however the Mekong delta is two hundred and fifty miles in length and the Red River in which delta both Hanoi and Haiphong are situated runs a hundred miles inland. Like the delta area of Bangladesh, the Vietnamese deltas, along with the Thai delta and river deltas in Egypt, Russia, Iraq, the Netherlands, Louisiana, California, Brazil, well, a hundred places, are all prime danger areas for ocean level flooding. It is just a fact of human civilization that we like to build big cities and intensive agricultural developments in river deltas, most of which are only scant feet above sea-level. In the case of Vietnam, when Saigon and Hanoi, plus Hue, Da Nang, Haiphong and Cam Ranh are flooded that there won't be anything left but jungle hillsides. Which is why we bring this up. If the prospect of coastal inundation is released too soon, militant nations such as Vietnam could very easily turn to conquest of neighboring highlands, such as Laos, to find dry ground. If they find out about this prematurely, or if it fails to occur, the resultant warfare could be horrendous and unnecessary."

Andy could not believe his ears, the CIA seemed to be considering an effort to keep the inundation information secret to keep foreign countries from taking military advantage of the situation. What about the number of lives that could be saved with proper warning?

The international briefing finished up with estimations of numbers of refugees, loss of agricultural land, loss of industrial base and financial effects on the major world powers. With the exception of the one incredulous admiral, not one of the audience from the President on down had interrupted the Central Intelligence Agency's predictions of impending utter chaos for the world. They were awestruck, as was Andy. Even though he had contemplated the dire consequences for months now, he had never fully appreciated the incredible facts of what was facing the nations of the world. And especially in light of the fact that the danger might be just years away and not decades or centuries.

The Director of the Federal Emergency Management Agency (FEMA) came up next for an analysis of the effects on the U.S. populace. FEMA had obviously done its homework well. As directed by the President in March, they had created contingency plans and situational analysis of the coastal inundation threat for the US. They had evacuation scenarios, refugee resettlement plans, financial and industrial predictions and even educational and public relations plans.

Although obviously the fruit of efforts of many agencies and merely coordinated by FEMA, their work was excellent. However, as he listened to the presentation, Senator Knowles could not help but wonder why America had not done as much to avoid the crisis as it did to plan for the damage. If the American leadership in a former administration had cooperated in giving some clout to the Rio and Kyoto Treaties and followed up with other preventative actions when they had the chance, they might not have had to protect the cities from the ravages brought on by the increased carbon dioxide and its greenhouse effect which the organizers of the Rio and Kyoto Conferences tried to prevent. If Congress had not drug its feet in restricting CFC production, the President might not need to consider plans to shelter and feed twenty to thirty million American refugees. With the wisdom of hindsight, Andy Knowles could be critical of the handling of the environment. On the other hand, Andy thought, if Andrew Knowles had been elected President years ago instead of the President's predecessor and had the burden of the struggling economy on his hands, would he have been able to push the environmental reforms through? Heaven knows the Congress was being tough enough on them this year, even with the newfound GAIA data and the reports of environmental degradation they had heard.

When the FEMA briefing was finished Dr. Fairchild started to go back to the podium, but the President, with a wave of his arm, indicated for him to stand aside. Instead, the President, without speaking, pulled himself out of the center chair. Andy noticed how the President used his hands on both chair arms to lift himself up; the President was obviously tired.

From the podium, the President surveyed the assembled personnel. Andy thought he could see the President pause for a moment when he saw Andy, but Andy could be mistaken. The President spoke, his voice was hoarse as it often was, "This ran a little longer than I thought it would. I need to get out to Andrews

196 Kevin E. Ready

for my flight now. I understand there will be a bit more for you to see and hear. Before I go I want to say something though.....I wanted everyone to hear it from me, personally." He paused a moment, clearing his throat. Andy was almost certain that the President was glancing directly at him.

"I want everyone to know that I don't want a repeat of Miami on this. We apparently don't have a handle on the timing of this thing and we aren't even sure what "this thing" is. We are going to pull out all of the stops to get a handle on what's happening. If we can find out if this ice is going to let go, we will. If we can find out when, we will. We will also do whatever we can within our government to get ready for any eventuality. However, I am not going to be responsible for creating chaos when we don't know what we are dealing with. Antarctica has been there for a million years and I'm to going to let anyone go off half-cocked and say its floating away until we know for sure. We have seen that an ill-timed word can cause riots and financial chaos and the CIA has told us it could cause wars. So, for now, consider yourselves and your people under a direct Presidential order not to divulge this ice pack thing until I say otherwise. I trust that is clear enough.

"Now, I want everyone to get on the bandwagon and help get this picture cleared up for us and to help get ready for what might happen. Thank you."

Everyone stood as the President left. On his way out the door the President turned and whispered to a Marine major by the door, who nodded. After the president and the staff who left with him had gone, the major came up the side aisle to where Andy sat.

The major bent over and whispered to Andy, "Senator, the President would like to see you."

The major turned away and Andy followed. Going down the aisle, out of the corner of his eye he saw the Vice-President watching, with her brow slightly creased. Andy locked eyes with her and raised his eyebrows to indicate he did not know what was going on.

The major handed the senator off to a Secret Service agent at the Situation Room check-in desk. The agent led Andy to the left, toward the main White House. They went up another flight of stairs and Andy found himself confronted by the First Family, personal staff and a pile of luggage. The President was talking to his wife.

Seeing Andy, the President broke away from his wife and came toward Andy. He motioned for Andy to follow him into a room. As he followed Andy waved a courteous hello to the First Lady, who smiled back.

The room was a formal anteroom. The President stopped and turned to Andy, reaching out to shake his hand.

"Andy, I haven't spoken to you since your wife.., the storm. How is she?"

"She is doing well. Fully recovered. Turned out it was just a concussion. She came out of the coma as soon as the swelling went down. I think she got a call from your wife in the hospital."

The President seemed mildly surprised at this, but the expression passed and he continued, "Andy, I wanted to talk to you about what I said down there." The President looked Andy in the eye. "I know how hard it would be for someone in your position to keep quiet about the flooding issue. I just wanted to tell you that I think it is best for all of us to keep this thing to ourselves until we know what is really happening. You have my word that my administration will do everything possible to forestall this, this disaster, and that we are doing everything we can to push the environmental package. We just don't want any more unnecessary events like the Miami riots to happen."

Andy waited to speak. He thought over the alternatives and decided, "You have my word I won't go public without talking to you first. Is that OK?"

The President smiled and nodded, "I guess that is all I could hope for." They turned and walked into the hall. The baggage was being carried out by the White House staff. Andy's escort was still waiting, as was the First Lady down the hall.

The President shook Andy's hand at the door to the ante-room.

"Thank you, Senator."

"Have a good trip, Mr. President. Don't let the Japanese sell you anything while you're over there."

Andy was still too far from the First Lady to say anything so he again just waived again and turned to follow the agent back to the Situation Room.

———

September - Ross Ice Shelf, Antarctica

John Colton was struggling to hold on to the dream. Becky and a lovely little girl (his daughter?) were cavorting on a beach and he was very determined not to leave them. But reality soon won out.

One of the Navy technicians was banging a coffee cup on one of the metal lockers, trying to wake him with the noise.

"Dr. Colton, they want you in the radio shack," he was saying. "Dr. Colton, get up!"

"OK, OK, I'll be there in a minute."

Colton rose to sitting position on the cot. He yawned and tried to stretch the kinks out of his back. He was still in a full set of thermal underwear, so all he had to do to get dressed was slide into the one piece snow suit and slip on boots. Of course, he would add gloves and headgear before running across to the communications shelter. The Navy men who regularly manned this outpost never used the full outfit to run between the shelters, but Colton saw no purpose in such showings of machismo. But then the women crew members did it also; was it proper to call it machismo? What was the feminine form for machismo? Machisma? Feminismo? It did not matter, he would wear the full cold weather gear.

As he struggled to get the nylon clasps on the boots to latch, Colton looked around him. He had only been there two weeks, but already he seemed to be getting used to the place. Except for the piles of cold weather gear all around, the living quarters of the Navy encampment on the Ross Ice Shelf was much like any military barracks anywhere. Cots, lockers, pictures of home and loved ones, posters.

The posters were what gave the place its distinction. A poster for the Pittsburgh Penguins hockey team seemed right at home, but was right next to a Club Mediterranee poster of a girl sunbathing on a black sand beach. The most telling poster was all by itself on the opposite wall; the old poster "NAVY - It's not just a job, it's an adventure" seemed particularly apropos.

Colton's former tormentor with the coffee cup appeared again. "Radio says they have an urgent call-up for you on the SATCOM network. They say it's Washington and Boulder on a conference call."

Colton nodded. He did not really believe it. Washington and Boulder were waking him up for a conference call at the farthest end of the earth. Of course, you could forgive them for calling in the middle of the night when night was over twenty hours long.

John Colton went to the door. He paused for a moment and tightened the velcro tabs on the gloves and head wrap. He had learned a healthy respect for the cold down here. After taking a last breath of warm air as he opened the first of the double airlock doors outside, He braced himself and pushed out into the bitter cold.

When the cold hit him, he was thankful for the plastic tube between the buildings. It was lit by a string of dim electric bulbs and only kept the frozen winds out, but the inflatable tube system kept regular movement between buildings possible in the sub-zero air. Although in the shelter of the tube, Colton ran for the other end. Even in the tunnel the cold stung his face. It was not as bad as the run in from the plane two weeks before, but nevertheless it was cold. He had actually seen the Navy men run between buildings in skivvies, but he realized that was probably more of that ever-present machismo for his, an outsider's, benefit.

Colton passed through the other airlock and into the communications and operations building. Captain Daniel Martin, the detachment commander, was there waiting for him along with an enlisted man.

"Satcom is over here. You can have it on headphone or speaker. Just flip this for speaker, we'll sign on for you," Martin said.

"Let's put it on speaker. You probably should be in on this."

The enlisted man flipped the toggle switch and pushed two chairs over to the bench the radio equipment was on. Colton and the navy commander sat down.

"Golden Eagle this is White Maiden. Golden Eagle this is White Maiden." The radioman said into the microphone as he pushed a foot switch.

"Golden Eagle, over."

"This is White Maiden. We are ready for traffic."

"Roger, stand by for encryption synch on my mark. Five - four - three - two - one - mark." On the mark the radioman flipped another switch and a short burst of crackling white noise issued from the speaker, then clear silence.

"Doctor Colton?" a voice asked from the speaker.

"Yes." Colton answered.

The radioman pushed the foot switch toward Colton and said, "Push down to speak and let up to listen."

Colton nodded and pushed the switch down.

"Yes, this is Colton."

"Colton, this is James Fairchild, at the White House. We have your colleagues in Boulder on the circuit. Dr. Brady?"

"We're here. John. We have Midge, Bart and the rest of the staff on speaker phone here. How are you doing?"

"I'm fine. Not quite adjusted to the constant darkness and sub-zero temperatures, but OK."

"Before we get on with business, I have a short message for you. Rebecca sends her love and tells us that she is just fine and the baby is still doing well and on schedule. You'd better finish up there and get back before the baby arrives without you here. She decided to visit her parents while you were away. She is in Palo Alto."

"That's a good idea. Give her my love."

"I will."

Fairchild's voice came on the line. "How are things going there? How about the new equipment?"

Colton answered, "The Navy had things well in hand when I got here. We hit bottom on the hole at our location last night, and we hit open water and slush before we hit bedrock."

He paused for that information to sink in. "They have the six outlying stations set up and will be drilling in the next few days. The echo sounding equipment is being set up now and we should have some sonar views of the bottom soon. We are still not sure how useful that will be in finding the bottom of the icepack, but it will show the ocean bed and that will give me a chance to do some recalculations of the ice pack volume."

Brady's voice could now be heard. "NASA has the GeoSat XI satellite reoriented and programmed to give us a precise altitude mapping of the surface of the Ross and the Ronne. They say they can get within five feet of elevation. That should finish the equation about how much ocean displacement the ice shelves could cause."

Colton spoke again, "I had a chance to speak with the British and New Zealand seismologists visiting at McMurdo when I changed aircraft. Now that they have an idea for exactly what we need, I think they can give us a better idea about the extent of the ice-pack shifting seismic events."

He continued, "The Navy guys, and gals, are really great. They have a good grasp of the data format Scripps needs to feed the icepack data into the GAIA system. I still want to be here when the outlying monitoring stations come on line. I should be out of here in two weeks, three at the outside. Oh, Dr. Fairchild. Those oilmen you kidnapped from Exxon in Alaska are working out well. They know the climate and had no problem getting the drilling and echo sounding going out here. Couldn't have done it without them."

Fairchild spoke again, "I am glad that worked out. Well, I guess that's it for now. Give the President's regards to everyone. We are all quite proud of the spirit and the speed that this project had been carried off with. Out here."

"Out here." Colton said as he handed the mike to the radioman.

———

September - Portland, Oregon

The crowd was loud and raucous. Those lumbermen who rose from the lumber camps of the Pacific Northwest to management status with lumber or paper companies, or with lumbermens' unions, kept a modicum of their lumberjack's demeanor, especially at an industry convention. And with this dinner following directly after a two hour cocktail party and reception, even the more reserved lumber and forest products executives were getting boisterous.

As a veteran of many a stump speech in loud and raucous union halls as well as a couple decades of being invited back to this gathering, Senator Harold Hudspeth knew how to handle a crowd like this. He was in his element.

Dinner had already been served. Brown slabs of meat, a dollop of mashed potatoes and green beans. Hudspeth had eaten the same meal a thousand times in his political career.

Hudspeth's dinner had been interrupted a half dozen times, but he did not mind. It was interruptions like these that made his political machine work.

Each interruption started with a tap on the shoulder or a cleared throat behind him. "Senator?"

He would turn and smile at the person like he was meeting an old friend, whether he recognized the person or not. A good politician never let on that he did not recognize or remember a constituent or contributor. A good politician, and that was what Harold Hudspeth considered himself, had a hundred different ways to greet someone without having to say their name since it was impossible to remember the number of names that people expected. "Well, hello there. How are you doing?"

"Fine, Senator. I just wanted to let you know how appreciative we all are up in Okanogan County. You're doing a great job." An envelope appeared. "We just wanted to give your campaign committee something to help out."

Hudspeth would palm the envelope with his left hand and in the same movement pump the offering hand with a rough, but warm handshake that only came from years of practice.

"Thank you very much. I appreciate your support." Hudspeth would say.

Most of the checks from a crowd like this would be for a thousand, the highest amount allowed from individuals. A few would be five hundred. However, a few groups represented here, trade organizations or unions, had organized political action committees (PACs) and their checks would be a uniform five thousand dollars each. The eight or ten envelopes and checks Hudspeth would have in his pocket at the end of an evening like this could contain anywhere from ten to forty thousand dollars or more. A few envelopes, from the old pros and PACs, would have a check both for Hudspeth's re-election campaign and for his party's national Senate Campaign Committee which could be earmarked by the Committee to be routed back to Hudspeth or used on his behalf.

The president of the Pacific Forest Products Council finally took the podium and started a belabored introduction of the guest of honor and featured speaker, Hudspeth. The introduction was basically the standard bio information supplied by Hudspeth's senate office, but was interspersed with testimonials of the good senator's undying support for the Council's objectives and the lumber and forest products industry, not to mention jobs and profits.

Harold Hudspeth rose and walked over to the podium, smoothing lumps and wrinkles in his suit with a quick jerk of his jacket front as he walked. He shook the hand of his introducer and waited for the applause to die down.

"Thank you very much for the kind words, Ray. I guess I did well enough in Washington this term to be invited back here again." Everyone present was aware of what Senator Hudspeth had done for the lumber industry, not only that term, but for many years. His work in trying to deregulate the timber industry and his partial success in taking the bite out of the Endangered Species Act were well known. So, when he said this and smiled, the crowd smiled and applauded again.

Senator Hudspeth started to speak again, but instead of words, the only sound heard by the audience, and for most the last sound ever, was the thunderous roar that accompanied the instantaneous disintegration and incineration of the podium, the dais, Senator Harold Hudspeth and the honored guests of the Pacific Forest Products Council.

The FBI report would indicate that the cause of the explosion was five sticks of dynamite, of the type used for road clearing by the timber companies. A case of the explosive had been reported missing by a logging company in northern California following a "monkey-wrenching" attack by radical environmentalists, calling themselves "deep ecologists" or "ecoteurs." The stolen dynamite had previously shown up in attacks on logging company equipment and facilities in California and Oregon, as well as an aborted attempt to blow up Hoover Dam. None of these previous monkey-wrenching incidents had killed anyone. But this time was different.

The initial force of the explosion was enough to smash Senator Hudspeth and the others on the dais against the ceiling and kill many in the front rows of the audience. But what killed the large numbers of people in the audience was the subsequent inferno.

The laboratory tests conducted by the FBI in their investigation of the assassination of a member of Congress, showed that the five sticks of dynamite were set off inside two five gallon commercial floor wax containers hidden under the dais platform. The contents of the floor wax containers was a mixture of gasoline and powdered laundry detergent in exactly the same proportions as the 1960's radical publication The Anarchist's Cookbook had given as it's recipe for napalm. The force of the explosion had splattered the home-made napalm jelly throughout the hotel auditorium. In the next flashing instant, all of the two hundred odd people who had heard Hudspeth's opening sentence were engulfed in the inferno.

Within minutes of the blast, a call came in to the San Francisco Examiner national news desk. A junior editor on the night shift answered.

"Examiner, National Desk."

"I'd suggest you record this," a woman's voice said.

"Wha'...?"

"I am going to make an important statement and you should record it. Can you handle that or should I call another paper?" The arrogance of her voice did little to hide the nervousness.

"OK. Got it. Go ahead." There was a beep on the line.

"A few moments ago a great enemy of Mother Earth, along with many of his despicable handmaidens, was eliminated from our biosphere. Senator Hudspeth represented the worst of the enemies of nature and life on our planet. His only concern was corporate profit, and economic and political power. He would sacrifice the very essence of life on our planet to achieve those goals.

"The Friends of the Mother will not allow the likes of Senator Hudspeth to kill all those things we cherish. We hereby declare war on those who would defile the earth. Those who refuse to nurture and cherish the gift of life are on notice that we will not allow the gift of life to be taken from the rest of the natural world. Be warned!"

"Just exactly what...."

"That is all." Click. The line was dead.

————

September - Boone, Iowa

Jared Tolliver stared blankly at the check. The co-op elevator manager was speaking, but Jared did not hear what he said.

"Pardon. What was that?" Jared finally asked.

"I said, Sorry it isn't any more. At least the market price being so high helped out a little." The manager repeated as he closed the check ledger.

"I guess I should be thankful I got anything."

"If you didn't you wouldn't be alone. I'd say there is a good twenty farmers who regularly store here who didn't get any crop at all."

"Well, next stop is the bank. They're holding my crop insurance and subsidy checks and I'll have to dicker and see what I can get them to let me keep from that. Too bad you can't get crop insurance that pays you what you lost on the final market price rather than the spring price support levels. Heh?"

"Shit. If they did that farming would be worthwhile. Never happen."

"Yeh, I suppose. See you at the game. Your boy playing again this year?"

"Sure is. Can't wait."

The chime on the door clanged as Jared left the grain elevator office. As he walked away he carefully folded the check and slid it into his bib overall pocket. He hopped into his dusty green Chevy pickup and pulled onto the highway.

———

Jared looked and felt out of place in the leather chair in the bank lobby. As he waited he looked down at the floor where his scruffy, stained steel-toed boots contrasted with the thick plush carpet and mahogany magazine table.

It was the same bank lobby he had come into with his father decades before. Only then it had been a local bank and his father had introduced him to the president, a personal friend. Now the bank was a branch of a multistate bank holding company. Jared recognized a few of the people in the lobby, but he was unfamiliar with the name, Charles M. Nelson, Jr., of the Managing Vice-President outside whose office the receptionist had told him to wait.

The door to Nelson's office opened and slammed shut. Jared looked up to see a young farmer, who Jared recognized as the father of one of his son's friends in the Boone 4H Club, storm out of the office. Jared raised his hand in greeting, but the man did not notice him. He was staring straight ahead with a lock-jawed expression that was a mixture of anger and desperation. He carried a sheet of pink paper in one hand and his baseball cap in the other. The young farmer walked quickly to the front door and with a stiff arm slammed the heavy glass door open ahead of him and walked out.

Jared was still looking towards the door when the receptionist's phone beeped. "Mr. Tolliver, Mr. Nelson can see you now."

Jared got up and entered the office. A thin, bald man in a dark blue-gray suit was putting a folder in a file cabinet and taking another out. He indicated with one hand that Jared should come in.

"Good morning, Mr. Tolliver. I don't think we've met before." Nelson came over and shook Jared's hand and said, "Please, have a seat."

Jared sat and said, "I didn't know Marv Franklin had been replaced."

"When the banks merged last winter, they decided to move the managers around. This summer I was sent in from Illinois to Boone and Mr. Franklin is in Wisconsin. They do that so the bank's staff is integrated throughout the company."

"Marv was here a long time. He gave me my first car loan, must be twenty-five, thirty years ago."

"Hmm, interesting." Nelson said, but he did not sound interested.

"I'm here to talk about my loan and the crop insurance and subsidy payments that you have for me."

"Yes, I'm glad you came in. I would have been calling you anyway." The banker was thumbing through the file.

"We ought to work out what you're gonna roll over for next year and how much of the federal money you need." Jared said.

"Of course, but I don't think there's much to work out." He glanced up at Jared and then back down to the file. "Our home office has established a policy on these types of matters and I'm afraid there is not much I can do about it."

Nelson paused and finally looked up to meet Jared's eyes. "Your land and equipment loans are secured so they will, of course, continue. We will pay the outstanding interest and the principal due on those loans from the crop insurance and subsidy money. Likewise for the FMHA guaranteed loan, but that's rather small. Unfortunately, I have had to apply the bulk of your funds to the unsecured operating capital loans."

"What do you mean you've had to apply my proceeds. Where do you get off doing anything without talking it over with me?"

"Your loan agreement gives us a power of attorney on the federal checks, that's why they send us the check."

Jared took a deep breath to control his anger and apprehension, "And you're not going to roll over any of the operating loan. Or give me a new loan."

"The bank has not made a decision on next year's operating loans yet. It all depends on the status of crop insurance next year. The weather problems and the uncertainty of the crops recently have made banking almost as hazardous financially as farming. We won't know for sure until mid winter what our policy is on future farm operating loans."

"So what's left after you take the operating money out."

Nelson ripped the back copy off of a multi-page form in the file folder. It was a pink sheet like the young farmer had carried out. He handed the form to Jared.

"This shows what application has been made with your funds. The interest payments and pay-offs are listed in the middle.

The amount in the box at the bottom has been credited to your checking account."

Jared was looking at the form, barely comprehending. His eyes caught on the box at the bottom.

"Damn. That isn't even enough to live on through the winter, let alone farm on in the spring."

"Mr. Tolliver, there is a very good chance that once we are able to get a good perspective on the outlook for next year that we may be able to advance you operating funds as we have in the past, or at least a sizeable percentage of what we always have. It appears from the subsidy invoice that you did reasonably well compared to most. We simply are not in a position to advance next year's farm operating money at this time or roll anything over as we have in the past. Things are too uncertain after four, five years of bad crops and with this whole climate problem hanging over us. We should be able to know more in January."

"And what should I tell the Feed and Grain that wants a down payment on the spring seed order now. `I should be able to get a good perspective on the down payment by January?'"

"Believe me, Mr. Tolliver, I understand. Everyone else is in the same shape."

"Right. That's very comforting."

Jared left the office. He did not slam doors like the young farmer had before him, but he suspected that his expression looked much the same.

———

September - St. Petersburg, Russia

This was the way it usually worked. If you had a warm, early spring, then autumn would be short and winter would come early. But not this early.

The summer had been unusually cool. The warm days of August which the denizens of northern Russia dreamed about all winter had been rare. The thermometer had risen into the eighties only a few days in August. Nikolasha Litasku had entered each day's readings in his personal journals. Quite often the entry had included the description "Overcast, brisk west wind, unseasonably cool."

His granddaughter, Daine, had been forced to wear a winter coat on the first day back to school in September. His daughter-in-law had left Nikolasha specific instructions as to this when she and Istvan had left for work that morning. Nikolasha had bundled Daine up and walked her to school. He bundled himself up, too. His eighty-five year old bones did not take kindly to the cold air like they once had.

By the time school was out that afternoon he was glad he had followed Sofia's instruction. As he waited in front of the school for the children to come out the first drops of freezing rain started to fall.

By evening, a full scale icestorm was in progress. This fact was studiously recorded in Nikolasha's journals. It was also taken note of a half a world away in the GAIA offices in Boulder, Colorado. They also noted that the entire Baltic and Scandinavian region was locked in the earliest winter storm on record and that the temperatures were the coldest ever for northern Europe in the first week of September.

October - New London, Connecticut

VIA NAVCOMMFAC - 2 PAGES XMITED
S E C R E T (XGDS-2)
FM: CMDR COHEN, USCG, NEW LONDON, CT
TO: GAIA OFFICE, NCAR, BOULDER, CO
INFO: SCRIPPS INST,(ATTN: DR. COLTON) SAN
DIEGO, CA
GAIA DISTRIBUTION LIST ECHO
NSC, WHITE HOUSE, WASHINGTON, DC
SUBJ: SITREP OF SIGNIFICANT MARITIME
INFORMATION - SEPTEMBER
1. USCG FISHERIES ENFORCEMENT OFFICE
REPORTS SIGNIFICANT REDUCTION IN NORTH
ATLANTIC FISH CATCH JULY - AUGUST.
COINCIDES WITH REPORTS FROM GAIA
SOURCES OF LOWER SALINITY, HIGHER WATER
TEMPERATURE AND INCREASED STORMS IN NORTH
ATLANTIC AREA.
2. ICEBERG SIGHTINGS CONTINUE TO EXCEED
ALL RECORDS, IN BOTH SIZE AND FREQUENCY.
NORWEGIAN COAST GUARD AUTHORITIES REPORT
THAT DESPITE INCREASE IN ICEBERGS IN
GREENLAND AREA THAT SEA ICE REMAINED IN
PLACE NEAR SPITZBERGEN FOR ENTIRE SUMMER
PERIOD. THIS IS POSSIBLY INDICATIVE OF
CHANGING WARM OCEAN CURRENTS, RECOMMEND
CONFIRMATION FROM SCRIPPS.
3. COLLATERAL INFORMATION - INTERNATIONAL
MARINE MAMMALS CENTER REPORTS SEVERAL
TRADITIONAL WHALE WINTERING AREAS IN
SOUTHERN HEMISPHERE WERE UNUSED THIS
SUMMER. RADIO TAGGED BLUE AND BALEEN

WHALE PODS APPARENTLY WINTERING IN HIGHER
LATITUDES. THIS IS POSSIBLY A REACTION
TO INCREASED SEAWATER TEMPERATURE IN
SUBTROPICAL AREAS. (SEE BELOW)
4. USN, CINCLANT REPORTS THAT IN MID-
AUGUST A SQUADROM OF U.S. NAVY SHIPS
(DESRON SIX) ON STATION IN PERSIAN GULF
HAD TO MAKE UNSCHEDULED PORT VISIT IN
BAHRAIN DUE TO LACK OF FRESH WATER.
CONVENTIONAL NAVY VACUUM CONDENSER TYPE
FRESH WATER EQUIPMENT REQUIRES RELATIVELY
COOL SEA WATER INJECTION TEMPERATURE
TO ALLOW CONDENSATION PROCESS TO WORK.
HIGH TEMPERATURE OF PERSIAN GULF WATERS
REQUIRED SHIPS TO MAKE PORT TO GET FRESH
WATER.
5. THIS REPORT IS CLASSIFIED IN
ACCORDANCE WITH EXECUTIVE ORDER 14708
AND INFORMATION IN THIS REPORT MAY NOT
BE RELEASED TO THIRD PARTIES WITHOUT
AUTHORITY OF NSC, WASH, DC.
XGDS-2
XXX

October - Bethesda, Maryland

Andy Knowles was in his living room preparing to delve into a stack of correspondence and papers he had brought from the office when the doorbell rang.

"I'll get it." Molly said from the hall.

Andy continued to read until Molly spoke from the doorway. "Andy, there are some gentlemen to see you."

Andy turned in his seat to see two men in business suits standing behind Molly in the hallway. Molly backed away and the older of the two men stepped forward, flipped open a wallet showing an official I.D. and said, "Senator, I'm Special Agent Jay Tierney, F.B.I., and this is Special Agent Gabriel Marquez, Secret Service. I'm sorry to bother you at home, but there is an urgent matter we needed to speak with you about."

Andy looked to Molly and then back to Tierney, "Certainly, come on in." He turned and indicated they should sit.

Molly slid the living room door closed from the outside. The men took their seats in the living room. Andy set his paperwork aside and sat down to face them.

"I didn't know FBI teamed up with Secret Service." Andy said, as though a question.

"We do when we have a common interest or mutual need. Again we apologize for interrupting you at home, Senator, but as you will see we had reason to want to speak directly to you and avoid your staff." Tierney said as he turned to the other man and held out his hand. Marquez handed him a file folder, which Tierney opened. He handed a sheet to Andy.

"Senator, this paper is a computer reconstruction of a burned document. Do you recognize it?"

Andy took the paper and looked at it. He nodded his head. The embossed Senate logo was unmistakable. "Of course, this is a memo my administrative assistant gave me in May prior to a meeting with Senator Hudspeth. This burning, was it in the fire with the Senator, when he died?" Andy assumed the obvious and immediately wondered how Senator Hudspeth had gotten the memo Derek had written for Andy about Senator Hudspeth.

"No, no. Not in that fire. This was found in a fire last week. But before we get to that could you tell us more about the document."

Andy paused for a moment, then continued, "Yes, I suppose I can. It is a bit confidential, but I guess not so much, in light of Hudspeth's death. This was prepared for me prior to a meeting I had with Senator Hudspeth about the environmental programs which were before the Senate. His committee had been slow to pass on the funding bills I needed to get moving on for the President's..., the Nation's environmental recovery program. He was concerned about the forestry restrictions and wanted to meet to negotiate a solution."

"And about this paper. Who prepared it and where would it be filed?"

"As I said, my A.A. gave it to me. It would have been prepared by one of my staff for him. It is a working document, to prepare me for the meeting with Hudspeth. I remember that I read it on the way to the meeting with Hudspeth and Derek took it back. I was going right to the meeting, at lunch, so I wouldn't have taken anything with me. And actually it might not have been stored anywhere. If Derek didn't see a need for it after the meeting, which was, by the way, successful, he might have discarded it. Now, would you mind telling me where you got this?"

"One more question. If he had discarded it, what would have happened to it?"

"For obvious political reasons. Everything from our office is shredded and then recycled."

"Looks like this didn't make it?"

"Looks like."

"Well, Senator, last Tuesday a strike force of F.B.I. and Alcohol, Tobacco and Firearms agents, were closing in on an apartment building in Berkeley, California, but just as they were about to go in, the building was fire-bombed, some sort of incendiary device, possibly homemade napalm. They got the fires out and salvaged a number of documents and other evidence. They were investigating Hudspeth's death and were acting on a tip from a "Friends of the Mother" member who was arrested in Oregon."

Tierney continued, "They obviously got the right place. They didn't make any arrests, but it looks like the place in Berkeley was the staging area for several `Friends' members.

"This memo of yours was just one document they found. It was in a charred folder of materials about Senator Hudspeth. They had press clippings, schedules, campaign brochures; a pretty good dossier on Senator Hudspeth. Included was a pamphlet on the Pacific Forest Products Council convention where he died."

Andy raised his eyebrows and the special agent nodded in silent communication with the Senator.

"However, the investigation of Senator Hudspeth's death is no longer our greatest concern."

"Yes?" the senator asked.

"We are trying to find out where your document came from because it might shed some light on some other documents we found. Right behind the folder of information on Senator Hudspeth, which we definitely believe was used by his assassins, was an identical information dossier on the President and the Vice-President. And on you."

———

November - Boone, Iowa

Jared Tolliver kicked the clump of grass. By mid-November a clump of rye grass should be dead from the frost. This clump was not brown and dead, but green and healthy. The latest killing frost he could remember had been Halloween, the year before last. Now, it was nearly Thanksgiving and they still had not had a frost.

You would not think that you would miss something like the frost, being the harbinger of the cold winter months to come. But here he was, worrying about it. The lack of frost and unnaturally warm weather this fall had everything as screwed up as the mess with the rains had last spring.

In the garden the tomato vines had sprouted another crop of yellow buds; they would never bear tomatoes, but they were flowering nevertheless. And the perennial plants like rhubarb and asparagus were growing huge, not having the cold weather as a cue to shut down their growth for the winter.

Jared had heard that some of the grain elevators were having trouble with spoilage of stored corn and soy beans. The cold weather of autumn usually enhanced the drying process getting the last excess moisture out of the stored crops with the cold dry air. With this warm weather and the several thunderstorms that had replaced the autumn snowstorms the elevators which did not have the expensive dehumidifiers were having real problems. A little spoilage in an elevator could spread and ruin a lot of grain.

Looking up from the clump of rye grass, Jared surveyed his fields. He had followed the advice of the ag experts and had left the fields unplowed for the winter, rather than plowing the stubble up and letting the land lay fallow for the winter. The new idea was that by leaving the soil alone and unplowed it would retain more water through the winter and would be less prone to erosion. After the wet spring he had suffered through he was not sure he wanted to retain moisture, but in the long run it had been the dry summer that had ruined his crop, so he was willing to try anything that would help.

But Jared Tolliver had never counted on what he now saw in his fields. In the southwest quarter, with its poor drainage, his hybrid corn crop had never really had a chance to sprout between the deluge of the spring and the drought of the summer. It had been a total failure. Now, in the heat of the prolonged Indian Summer, punctuated by those damn thunderstorms, the unsprouted hybrid seeding in the southwest quarter were sprouting up in the unturned soil between the sparse stubble left from his attempts to harvest in September. It was not much, and to anyone but a keen eyed farmer it would have been unnoticed. But, to Jared Tolliver, it was clearly row after row of tiny corn seedlings pushing up into the November air.

Jared made a mental note to himself to call the extension agent or maybe that scientist from Ames about this. He had to get some answers as to what was going on.

———

November - Washington, D.C.

"Are you sure?" a grim faced Senator Knowles asked Agent Gabriel Marquez.

"Of course we aren't absolutely sure. But the pieces fit together. And, but for the directions from the top not to embarrass you politically until we are certain," He paused, thought and restated himself. "Rather, we have been told not to embarrass you politically and not to act until we are certain. Therefore, we are letting you know before we contact her."

"Can I hear what she has to say?" Knowles asked.

It was the F.B.I. agent, Jay Tierney, who answered. "We have no objection to that at this point. But depending on what happens, or is said, we may have to stop and do things by the book."

"Understood."

Andy Knowles touched a button on his phone and his receptionist's voice responded, "Yes, Senator?"

"Christie, would you have Miss Lindquist come into my office."

"She's in back, I'll ring her for you."

The two agents and the senator sat in silence until the door opened slightly accompanied by a knock.

"Come in."

Dee Dee Lindquist came through the door, hesitating when she saw someone meeting with the Senator. "You wanted me...?" She was dressed simply in a plain styled blouse and skirt. Her short haircut and lack of make-up made her look quite young and her thin stature completed the waifish appearance.

"Yes. Please come in." Andy said and with a wave of his arm he indicated one of the chairs in front of his desk. "Please have a seat."

When Dee Dee was seated, Andy continued, "These gentlemen have some questions for you."

Tierney introduced himself and Marquez, who had seated himself on the coach near the far wall while they were waiting for her to come in. When Dee Dee heard the identities of the men she looked nervously at Andy.

Tierney took a seat next to the girl. He opened a file folder on his lap. "Before we begin there is a formality we must dispense with. Please don't be alarmed." Tierney paused and then continued. "You have the absolute right to remain silent and not answer my questions. You have the right to consult an attorney before answering. If you cannot afford an attorney, one can be appointed for you at no expense. Anything you do say can be used against you in a court of law. Do you understand these rights and do you consent to answer my questions?"

Dee Dee had sat open mouthed as Tierney spoke. Now she looked to Andy questioningly. "What is this about?"

"Dee Dee I can't help you on this. All I can say is that if you don't have anything to hide, you should answer the questions and get this out of the way."

"But, I don't even know what it is about."

Tierney spoke now, "We just have some general questions about a couple of your acquaintances and their activities."

"Well, sure, I don't mind, ask whatever you want."

Tierney turned his pen up to a point. He checked his notes and looked to Dee Dee. "Do you know Joshua Corbell?"

"Of course. He is a professor at Berkeley. He's chairman of my doctoral committee."

"Could you explain?"

"I am a candidate for a Ph.D. at University of California in Political Science. Professor Corbell is the chair of a three person committee who evaluates my plan and program for my doctoral education and dissertation. Sort of my mentors."

"How long have you known Corbell? And how did he get to be your chair or mentor?"

"I was his teaching assistant when I was studying for my Master's. When I was accepted into the doctoral program it was only natural that he be my committee chair."

"And what contact do you have with him now, recently?"

"This internship with the Senator's office is an approved part of my program. The research I do here will go into my dissertation. As I put together parts and pieces of the research I check with him to make sure I am on track. I write to him every few weeks."

"And what is your research subject?"

"The Effect of Special Interests on the Legislative Process."

A muffled noise from Senator Knowles interrupted Tierney. Andy waved a hand to indicate that they should not pay any attention to him. He put a fist to his sternum as though he had heartburn. Tierney continued.

"Do you include anything in these letters?

"Yes, at times I include documents and support materials. It is needed to document the work during the internship."

Tierney mm-hmmed and wrote something.

"Ok, now. Do you know a person named Paul Hilliard?"

"Yes, he used to be my landlord in Berkeley and we dated a few times, several, aah maybe two years ago."

"When was the last time you had contact with Hilliard?"

"Over a year ago, before the start of the school year. I hadn't wanted to continue with any relationship, it was going nowhere, and living in the building he managed was becoming a problem. So, I moved in with a friend to get settled for my first year in the Ph.D. program."

"And you haven't had contact with Hilliard since."

"No, none."

"Do you know of any connection between Hilliard and Corbell?"

Dee Dee took a moment to think about this question, a look of recognition came across her face. "Yeh, I hadn't thought about it in a long time, but, now that you mention it, when I first came to Berkeley for my Master's and got the teaching assistantship under Professor Corbell, he put me in touch with Paul to find a place to stay. Berkeley is notorious for being hard to find living quarters. That's when I moved into the apartment and it was sometime later that Paul first asked me out."

"Do you know how Corbell knew Hilliard?"

"I can only guess."

"That would be fine."

"Well, Paul was a biology major, he dropped out before he got his undergraduate degree. He was always interested in ecology and the environment. And the same with Professor Corbell, it is one of his pet peeves if you will. I can only assume that they had some connection the environmental protest groups in Berkeley."

"But you don't know for sure?"

"No."

"And do you have any contacts with those groups?"

"No. I went to one meeting, Paul took me. They were putting together a protest of Alameda County putting condos on a wetlands. It was a bunch of throwbacks to the sixties, longhairs, you know. Not my type of people. I never went again."

"Do you know the term 'monkey-wrenching'?"

"Yes. It's the term for taking radical action to fight for the environment. Like blowing up power lines in Arizona or driving steel pikes into logging trees to prevent cutting." As Dee Dee answered she darted a quick glance to Andy, who was watching her closely.

"Do you know if either Hilliard and Corbell have any association with monkey-wrenching?"

"No, no I don't. I can't see the professor doing anything like that, but I wouldn't put it past Paul. He had a freaky streak in him, a single-mindedness. That is part of why I never let our relationship get anywhere."

"Do you know anyone involved in any monkey-wrenching operations?"

"No." Dee Dee was slightly emphatic in her answer.

"Alright. Now," Tierney pulled a paper from his folder. "Do you recognize this?" He handed the paper to Dee Dee.

When she got the paper she looked at it. She closed her eyes for a second and then she looked to Andy. She let out her breath and responded to Tierney, "This is a copy of a list of bills on the environment in the budget committee that I prepared for the Senator on Derek's orders last spring. and, before you ask, Yes, it is one of the papers I sent along to Corbell."

"How did this come to be sent to him?"

"Like I said, it was expected that I document things and Hudspeth's work for the special interests is notorious. When I talked to Corbell in early May he suggested this might be a good case study, so I gathered what I could."

"Did you bring up Hudspeth's name or was it Corbell?"

"I had heard Derek, Derek Shaw, the Senator's A.A., talk about how Hudspeth was tying up the environmental programs in the budget process and I mentioned it to Corbell. Then he suggested that I put together what I could on Hudspeth for later research."

"And you sent that research dossier to Corbell?"

"Yes. We filed all of my stuff with my other dissertation materials at the department on campus. I get to keep my dinky office there even though I am in Washington on the internship."

"And Corbell had access to your office and those materials?"

"Certainly."

Tierney looked to Marquez, who nodded back.

"Thank you very much, Miss Lindquist." Tierney said as he closed the folder.

"Do I get to know what this is about? I can guess, but I would like to hear it straight from you."

"I guess that would be appropriate. Paul Hilliard and three others have been indicted in absentia for the murder of Senator Hudspeth and the others at the forest products convention. None of them have been seen in some time. We were also investigating Joshua Corbell as a known member of the monkey-wrenching group Friends of the Mother along with the other four. The Hudspeth memo was found in an apartment building that had been occupied by the group. We knew of your connections with both Hilliard and Corbell and we knew that you were the preparer of the memo for Senator Knowles. We just couldn't tie up whether it was Hilliard or Corbell who had gotten the memo from you. Now. if your story checks out we can probably clear you and probably indict Corbell. If we find out you haven't been forthcoming or we need more information we will be back. You are instructed to talk to no one about this. Understand?"

"Yeh, sure."

Tierney and Marquez nodded thanks to the Senator and walked to the door. Dee Dee started to follow when Andy spoke, "Just a minute, we need to talk."

When the door had closed Andy looked at Dee Dee squarely and spoke firmly, "I certainly hope they can tie this thing up without dragging you, me or this office into this, but whatever comes of this I have something to say to you." He paused and continued. "Whether you are an outside intern or a regular employee, when you are working in this office you have standards which must be met. For you to have taken a confidential memo, on my letterhead, out of this office for inclusion in your research materials without my permission is a serious breach of ethics, both personal and professional, not to mention common sense. Further, I would think a Ph.D. candidate in Political Science would have more judgment than to let sensitive political matters be included in research papers. Just how did you think you would use those papers. Do you think my interoffice memos belong as appendices to your thesis."

"I'm sorry. I wasn't thinking."

"Obviously. Now, what I want you to do is this. The Thanksgiving break is coming up. I want you to take two weeks off and get your head straight on this. I'll contact Tierney and make sure that they have no objection to you gathering up your papers in Berkeley. If they don't, I want you to get that stuff, sort through it and get whatever is politically sensitive back to this office. No, wait, if you can, get everything back here and Derek can help you sort. I don't want anything else to blow up in my face of this, Understand?"

"Yes, Senator."

———

November - Bangladesh/East Bengal

Akim Darpheel had been taught in his politico-military studies class years ago that conventional wisdom would dictate that to subjugate a people and assimilate them into a different country you should take over the existing structure and exert control through that structure until changes could be made, but the new government of the Indian state of East Bengal had no such opportunity. There was no existing power structure. There was very little left of society. And, except for the need to stop tens of millions of refugees from flooding into neighboring parts of India, there was absolutely no reason why any government would want to assimilate what was left of Bangladesh. Except perhaps to avenge Mahatma Ghandi's one failure, the failure to keep India in one piece after independence.

Darpheel had his hands full. Majlis had appointed him a brevet colonel in Majlis' token Bengali hierarchy put in place by the Indian Army and put him in charge of the Assam/Garo region. Now, Darpheel was working to coordinate the Indian and international efforts to assist in food relief and other aid to the chaos that was now Bangladesh. His experience at organizing training and supplies for the Bangladeshi Army was useful. The Indians were keeping fairly close to their plan, exemplified by Majlis, of having the "East Bengalis" be in charge of the new Indian state.

Quist and Mullahwi had likewise been promoted and placed by Majlis on his staff. A small security force of Indian troops was left behind in their village to provide the protection the three had given the village. Majlis had other things in mind for these men he felt he could trust. Darpheel had left Latifa back home with his family as he went with Majlis with the second wave of Indian forces into Dhaka.

What they had found was indescribable. Things had been bad when they had left in the spring, but the intervening months had been horrific.

When the central government had broken down, all else had gone with it. There had been no law enforcement, no supplies, no medical care and nothing of the amenities or societal structure the city of millions had before the catastrophe. The were stories of how individuals, police officials and often the Moslem clergy had tried to rally the populace in getting things organized. None had succeeded. Even with the force of the Indian Army behind them and international help, the task would be daunting.

The population of Dhaka had dropped to a fraction of its former seven million, perhaps only a million. And those who were left were pitiful. The strongest and the smartest had left before the worst of the starvation, pestilence and bloodshed. Pressure was on the Indian Army and Majlis' government to get things straightened out so that the millions of refugees in the camps on the three borders India had shared with Bangladesh could be repatriated.

After several weeks of work with Majlis' staff in Dhaka, Akim Darpheel was sent back to the Garo hills with instructions on organizing a local government in the hill country, his home, and the neighboring lowlands and population centers. As one of the few elevated areas of the new East Bengali state, it had survived the storm and flooding better than most areas. Darpheel's assignment was to get agricultural production and the market system back in operation so that the Garo Hills and East Bengali's part of Assam could funnel foodstuffs in to Dhaka and the devastated parts of the country as the tens of millions of refugees were sent home by India.

"By what authority do you do this?" the richly dressed merchant demanded, his face livid with anger. He was held back from approaching Darpheel by two Indian soldiers pinning his elbows behind his back.

"If you had read the warrant, instead of tearing it up, you would know. The Military governor of East Bengal has decreed that all stores of grain be released for shipment to areas which need them." Darpheel did not look at the man, instead he watched as the barefoot work crew under guard of the Indian troops loaded bag after bag of grain from the merchant's warehouse onto lorries.

"But, it is my life, my wealth, you will ruin me."

"You will be paid for what we take."

"But only in filthy Indian rupees, worthless Hindi scrip. I demand gold."

Now Darpheel faced the man. "You are not in a position to demand anything. Many of the bags are marked as relief shipments from America and Europe. It is illegal to sell such grain. You are lucky we don't confiscate the lot and jail you for black-marketing. As it is we will only penalize you double for the illegal bags. Hopefully for your sake, you will break even when the final tally is done."

Darpheel was pleased with himself. He was getting this confiscation job done without letting on that one of his objectives was to keep the merchants in business, funded with Indian banknotes to keep the market system operating. He would leave the man with enough cash and stock to get back on his feet. Such were the instructions from the economist from Delhi as relayed through Majlis.

Turning to of the soldiers holding the merchant, Darpheel said, "If he gives you further trouble, throw him in Mymensingh Jail. I am sure the thieves there, serving their sentence for stealing bread for their families, will be interested in hearing his demand for gold for his warehouse of grain." At these words the merchant avoided Darpheel's eyes and was silent. Darpheel gave a wink to the Indian sergeant who was restraining the man. He would not go to jail this day.

Colonel Akim Darpheel got in the front seat of his Land Rover next to the driver. This was but one of many similar operations his men were carrying out today. His program to counter the widespread hoarding was bearing fruit.

"Go to the rail siding. Let's see how the loading is going," he told his driver.

As the vehicle threaded its way through the bustling streets of Mymensingh, Darpheel took stock of his life. It had only been six months since everything for him and his country had gone into chaos with the typhoon. His life had changed so much. He had been a hapless bureaucrat for a failing government, homeless refugee, a new groom, a new stepfather, a mercenary for his father's village and now he was one of perhaps a dozen men who, under Majlis and the Indian troops, ruled his former nation with an iron hand. A nation that had gone from having a population of 120 million down to perhaps 60 or 70 million, with both deaths and refugees leaving the country. And a nation which was no longer a nation, having been swallowed up by a desperate, fearful neighbor; desperate to stop the flow of refugees and fearful that the same might happen to it. Darpheel wondered what the next six months might bring.

———

November - Central Valley, California

Paul Hilliard sat on a hillock, the late afternoon sun on his back. He pushed his long dirty blond locks out of his eyes and up under the headband he wore. He adjusted the eyepieces for better focus. The war surplus binoculars he had picked up at the pawn shop were coming in handy. With them, from his vantage point, he could survey several miles of the Aqueduct.

The California Coastal Aqueduct was but a part of the State Water Project. Together with the main California Aqueduct which supplied water to Los Angeles and the multiple systems of canals and dikes, the Project constituted one of the largest water diversion projects on the planet. As the most recent addition to the Project this was, to the environmentalists, the most dangerous. For it represented the newest attack on the cherished wetlands of the Sacramento/San Joaquin River Delta. Further, since its sole purpose was to provide water for irrigation and urban use for the coastal cities to the south it epitomized but another aspect of the rampant growth in California which they feared and fought. Clearly, the Project and the Coastal Aqueduct, in particular, were antithetic to the principles of the Friends of the Mother.

Hilliard leaned back on the overstuffed frame backpack and viewed his objective. It appeared exactly as it had in the brochures they had received from the public information office in Sacramento. To his left the lower canal portion of the aqueduct with its banks full of water sparkled in the last rays of sunlight taking on the teal blue of the late afternoon sky. Next came the lower pumping station, encircled with fences and bristling with the high tension lines that powered it. As he watched, the shadows of dusk having filled the valley before him, the ring of mercury vapor lamps surrounding the pump house flicked on under the command of an unseen solar switch. In the binoculars, Hilliard could see the trucks belonging to the workmen at the pump house. He knew that there was a similar pumping station to the south, hidden up in the hills, which was on the receiving end of the water pumped from here.

From the rear of the pump station arose the four huge pipes, 'the risers' the brochure had called them. Big enough for a man to walk upright inside them, the foursome of huge silvery pipes climbed into the hills behind the station like crazy parallel Tinker Toys. At every hilltop, and every valley and turn they passed through the angular concrete structures which held them in place and supported the weight of millions of gallons of water in the pipes. These were his objective, as other similar structures were the goal of his compatriots elsewhere in the state this night.

They had thought carefully about this. The critical pump stations were more expensive to repair, but they were also manned, guarded, and lighted, making them more dangerous to approach. On the other hand, the piping supports sat alone on hilltops and dark valleys up and down the coast and across the state. And, as the State Water Project information booklet had discussed, the pumping and siphoning system was an absolutely critical part of the Project, and had taken years to engineer and build. Since the objective of the Friends of the Mother was to keep the monkey-wrench in the works of the Project for as long as possible, the concrete supports would make a perfect target.

With the fall of darkness across the hills, Paul Hilliard set out. He had chosen three hilltops and one obtuse angled hillside support as his targets for the four satchels of high explosive he carried. The Friends had been lucky to be able to procure a good quantity of high explosives, the good DuPont plastique which was capable of cracking the reinforced concrete. Ernie Villanueva had been given the stuff in Mexico by persons unknown to the Friends and unquestioned by them. The self-righteous rarely question their benefactors.

Hilliard calculated that he could cross the open hillsides and reach all four sites and still put some distance between himself and the explosions before the 2:00 A.M. deadline the Friends had set for M-Hour, Mother Hour, as they called the time of every attack the Friends of the Mother made.

As he stood up from sitting on the rock, Hilliard felt the pain of a muscle cramp from having sat too long. For a moment he thought that he was getting to old for this, nearly fifteen years of monkey-wrenching expeditions. He quickly rejected the thought. This was his life, it made everything else worthwhile. It gave life meaning, to be doing something to protect the earth. Paul Hilliard pulled the backpack on and set out around the ridge line to the first concrete block and its Tinker Toy burden.

———

November - St. Petersburg, Russia

Doctor Sofia Litasku was waiting in the treatment room when the orderly brought the young woman in. Sofia could have described the girl's face as cherubic, had it not been for the tears and pain in her eyes. The girl was well bundled up, as all St. Petersburgers were these days and she had both hands wrapped in towels and clutched tightly to her breast. The orderly had her sit on the table and he left.

Dr. Litasku looked at the chart and spoke, "How did this happen?"

As the girl answered, the doctor carefully unwrapped the girls hands, "I was at work, at a bakery, when I suddenly noticed how late it was getting. I looked out the window and saw the trolley coming. With the new winter schedule if I didn't make that run, I might not get all the way home in Pargolovo before the trolleys and trains stopped for the night. So I washed off my hands, grabbed my coat and ran for the trolley. I had my coat on, but not my gloves when I got to the trolley and, ouch!, and when I grabbed the rails to pull myself up I just stuck there tight to the railings. I guess my hands were still moist from washing them. Some people tried to help, but my skin would not come loose, they pulled some skin off. Finally the conductor had to stop and get some hot water from a building to melt my hands loose. I thought I would die."

When Sofia saw the state of the girl's hands she had her lie down with her hands at her sides rather than sit, so that the hands would not be visible to the patient. The girl tried to look but Sofia pushed her down again.

With a swab and some iodine bath Sofia carefully probed the palms and fingers of the girl's hands. There was blood from the ripped skin in the towels, but most of the bleeding had stopped now. Both hands were the fish belly gray color of frostbite that she had learned so well over the past month. The full layer of skin was missing from several fingers on the girl's left hand and part of the fleshy pad of the palm on the right hand was pulled off and hanging on a string of skin. The remainder of both hands had been thoroughly frozen to the metal railings. Hopefully, the conductor had gotten her off in time. If the freezing had gone below the epidermis she might lose the fingers or worse. Even if not, the young woman was in for a painful healing process and probable scarring.

Sofia repressed the urge to scold the girl for her foolishness. It would do no good and she recognized the urge as her own overreaction to the dozens of similar cases her staff had seen since the cold spell had started.

Russians were used to cold weather. Lots of cold weather. But the treacherously cold weather that was besieging St. Petersburg this winter was unlike anything other than the stories told of Siberian winters. The long, cold winters Sofia remembered from her childhood in Moscow had been nothing like this. The piles of snow she was used to had not fallen this year and had been replaced by incessant, bitter cold, often reaching negative 15 degrees Centigrade at night, sometimes worse. Cold that could crack steel machinery and freeze flesh at a touch.

Sofia dressed the injured girl's hands and informed her she would have to stay at the hospital until they could determine the extent of the damage from the cold.

Peering into the dark street from her office window in the hospital Dr. Sofia Litasku saw the black sedan pull to a stop. She couldn't recognize the car or its driver, but she could tell the lights of a Volvo and there could be only one black Volvo pulling up to the hospital at this hour of the evening. Her police escort had arrived.

She grabbed her things and pulled on her coat and then her hat and scarf. When she reached for her mittens, she thought of the girl she had admitted earlier in the evening. She would make a point of telling Daine about this, if she drilled this into Daine, her daughter would never make such a foolish mistake. It was certain that Sofia would never forget to protect herself from the cold.... Taking stock of her own thoughts, Sofia recognized her own reaction to the girl's trauma. She must keep herself objective or she would be of no use to anyone. She walked to the front door and bundled herself further into the scarf.

Taking a last warm breath of air, she opened the door and trotted to the car. In spite of her preparation, she felt the stinging of the cold on her nose and in her eyes.

The night was still and without the usual sounds of a city. Everyone was staying in. She was lucky her husband had decided to pick her up. A walk to the Metro would have been miserable and waiting for a few trolleys that were running could be deadly.

As had become his habit since the cold started, Istvan was having his driver take him home, picking Sofia up on the way. The young driver, who lived in the militsia barracks near the motor pool, would park the car in the protected garage and come get the militsia commander again the next morning. This practice was not simple elitism on Istvan's part, he had never before done this, the Litasku's had no place to park a vehicle out of the cold at home and many engine blocks had frozen and cracked in the cold this month when the cold exceeded the limits of the poor quality anti-freeze available and the owner had not drained the radiator each evening. Avoiding the problems and inefficiency of the alternative, Istvan simply now accepted the perquisite of office of any militsia chief and took the ride from the official car. Sofia did not mind the loss of the principles Istvan had formerly maintained.

Istvan had the rear door open for her when she got to the car. She hurried into the rear seat with him and slammed the door.

The interior of the Zil was toasty warm in comparison to the outside. Zils did have especially good heaters. They had to in Russia. Istvan grunted an order at the driver and they pulled away.

Sofia stretched over and kissed her husband on the cheek. He did not respond except for a look at her. He never liked to kiss or show affection in uniform or in front of his men. Sofia knew this, but never followed his wishes. She figured it was her prerogative to tease him thusly.

Sofia loosened her scarf and collar. "How was your day?"

Istvan slowly shook his head and grunted, "Fine, just fine, and you?"

"The usual. New babies. Dying old people. And now, more of the frostbite cases."

"I know about that."

"What do you mean?"

"Four school boys tried to take a short cut to school and cross the Neva on foot instead of over the bridge. A workman saw them go through the ice mid-river. We never did find two of them. And by the time the crews got to the other two they were so frozen that we didn't even bother bringing them into the hospital. My forensic officer declared them dead on the scene."

Sofia had thought of telling Istvan about the young girl, but saw that he was in no mood to hear her sad stories. She decided to change the subject.

"Anything new on the weather forecast? Any end in sight of this?"

Istvan now looked at her and smiled the lopsided smile that he did when he had bad news. "No, the official weather telex from Moscow is the same as the television reports. Continued cold for the rest of the week, but there may be a frontal system by the weekend that could bring snow and a break to the stagnant cold."

Sofia shook her head, "Its pretty bad when you have to hope for a blizzard to warm things up. I'm just wondering what happened to all that talk of global warming we heard all last spring."

Istvan raised his eyebrows and nodded also. He had wondered the same thing many times. He put an arm around his wife and pulled her close for the ride home.

———

November - Central Valley, California

Paul Hilliard pushed the button on his cell phone to check the time. The dim light revealed the time to be shortly after midnight. He had just placed the last of the three hilltop demolition charges and was now scrambling up the hillside to the concrete structure supporting the pipes on their way up the hillside. In order to climb the steep hillside the pipes took a peculiar angle at the concrete support before rising nearly straight up to the next summit. Even better, the section of piping before this support stretched above ground across the small canyon he was now crossing. If he could knock out this one support the entire elevated portion across the canyon and the near vertical section up the next mountain would collapse, taking out probably a thousand feet of pipe. But he would have to hurry if he was going to get up this next hillside, place the charge and get safely away before 2:00 A.M. when he had set everything to blow.

The going had not been easy. He had gotten the other three charges placed, but the distance between the between the targets was more difficult to cover than he had hoped. And this last hillside was not any easier.

The canyon floor was rocky and in the moonless night he stumbled frequently. As he started up the slope he discovered that the smooth appearance of the slope from afar was caused by the loose shale that had fallen away from the cliffs above. Climbing the slope was like climbing a sand dune, every step caused him to slide part way back down. By the time he reached solid rock he was out of breath.

The rock area was not as strenuous a climb as the loose shale, but more dangerous. The weathered shale was a precarious foothold and he stumbled often, gnashing his hands and knees on the rock.

He finally reached a road cut in the rock face made by the construction crews to reach the structure above. He wished he had seen it before. It would have made things easier.

The structure was essentially the same as the hilltop ones. The four huge pipes were held in place by steel frames which in turn were attached to large rubberized bases attached to more steel embedded in the concrete, apparently a shock absorber system for earthquakes. The poured reinforced concrete served as both a base for the structure, and also a cover. From a distance it appeared that the pipes went directly through the concrete.

As before, Hilliard placed the explosive charge on the downhill edge of the structure, up under the pipes against the concrete and steel frame. This seemed to be the weakest part.

With the charge in place, he took a drink of water from his canteen and before setting this last timer he took another look at his watch. He only had forty five minutes left.

He did not want to risk having the other charges go off before this one and possibly have response crews find the last charge before it went off. So he would have to set the timer for the same time.

Thinking now about the access road he had found, he assumed he could utilize it to get away from the immediate vicinity and then cut off the access road into the rougher country to the west were the car was parked. He set the timer for 2:00.

Compared to the rough journey he had experienced earlier the road was a dream. He could not see road itself well in the dark but as he hiked down the groove he grew more at ease. Even in the starlight it was an easy walk.

A hundred yards below the structure Paul Hilliard took a step and then felt nothing under his boot. He tried to rock back on his other foot, but it was too late. He fell forward off balance into a dark gully, a cut washed into the access road by the winter rains a year before.

He did not fall far, perhaps ten feet, but landed face first on the other side of the gully on hard shale. He slid another few yards and came to rest in the bottom of the gully. He moaned once and tried to raise himself up, before passing out, his blood spilling onto the California shale.

At precisely 2:00 A.M. on Thanksgiving morning the California State Water Project was rocked by eleven explosions. Investigators would find one charge set, but undetonated at an auxiliary pumping station in the San Bernardino Mountains.

The eleven explosions each severed vital pipelines in the pumping and siphoning system of the California Aqueduct, the Coastal Aqueduct and the Los Angeles Aqueduct. It was estimated it would take years to repair the damage.

The water that was not able to be sent to Los Angeles and the other south coast communities went on its natural way downstream. However, in the seventy years since the project had been started the ecosystems on both sides of the Sierras, from which the water had been drawn, had adjusted to the loss of the water. The glut of millions of acre feet of unexpected water into Mono Lake and the Sacramento Delta with its intricate system of levees and canals wreaked havoc. What wetlands had been preserved in the Delta were now flooded into extinction and the saline ecosystem of Mono Lake was devastated.

Los Angeles and the rest of Southern California now strained the remaining water source, the Colorado River Aqueduct, to its limits. Attempts by federal authorities, not to mention Arizona, to prevent California from using more than its allotments of precious Colorado River water fell on deaf ears when Southern California was literally dying of thirst. Taking a back seat in priorities to household uses, the farmers of the Imperial Valley, the Coachella Valley and the vital citrus growing regions of southern California found themselves out of the loop when it came to water. The government of Mexico threatened to take California to the World Court if it did not allow Mexico its share of Colorado River water.

Paul Hilliard had still been unconscious when the charge blew away the piping supports above him. He had been almost correct in his assessment of the structure. When the charge detonated, the cross-canyon pipes tumbled into the gorge along with the concrete block. However, the pipes continuing up the cliff stayed in place and when the explosion released the pressure head on the pumping/siphoning action of the pipes the hundreds of thousands of gallons of water in the pipes in the hills above the explosion reversed themselves and flowed downhill toward the break.

Much of the gushing water flowed out the pipes and onto the access road, then coursing down the gully in which he lay. The torrent of water washed tons of loose shale and gravel into the canyon below. By dawn's light there was no sign of Hilliard's final resting place beneath the new alluvial deposit that joined the gargantuan silver Tinker Toys which lay broken in the canyon bottom.

———

December - New London, Connecticut

VIA NAVCOMMFAC - 2 PAGES XMITED
S E C R E T (XGDS-2)
FM: CMDR COHEN, USCG, NEW LONDON, CT
TO: GAIA OFFICE, NCAR, BOULDER, CO
INFO: SCRIPPS INST,(ATTN: DR. COLTON) SAN
DIEGO, CA
GAIA DISTRIBUTION LIST ECHO
NSC, WHITE HOUSE, WASHINGTON, DC
SUBJ: SITREP OF SIGNIFICANT MARITIME
INFORMATION - NOVEMBER
1. AUSTRALIAN NAVAL AIR RECON FLIGHT
REPORTED MASSIVE ICE FLOE AT 104 E 61
S. ESTIMATED SIZE OF FLOE WAS 110 SQ
MI WITH HEIGHT OF 100-300 METERS ABOVE
OCEAN SURFACE, AND SEVERAL SMALLER FLOES
AND BERGS IN VICINITY. PROBABLE SOURCE -
SHACKLETON ICE SHELF ANTARCTICA.
2.COLLATERAL REPORT - RUSSIAN NAVAL BASES
IN WHITE SEA AND KOLA PENINSULA AREA WERE
ICE BOUND AS OF NOVEMBER 11. THIS WAS THE
EARLIEST REPORTED CLOSING OF THE NORTHERN
RUSSIAN PORTS SINCE 1943. AS OF NOVEMBER
29 RUSSIAN NAVY CEASED ATTEMPTS TO FREE
CHANNEL WITH ICEBREAKERS DUE TO THICKNESS
OF ICE.
3. CHINESE MARITIME COMMERCE OFFICE
ISSUED "NOTICE TO MARINERS" WARNING OF
DANGERS OF RUNNING AGROUND IN APPROACHES
TO SHANGHAI WITH INDICATION THAT RECENT
FLOODINGS HAD CAUSED CLOSURE OF MAIN
CHANNEL DUE TO SILTATION.

4.INDIAN NAVY "NOTICE TO MARINERS"
REPORTS CONTINUED PIRACY ACTIVITY IN
FORMER BANGLADESHI COASTAL REGION. INDIAN
NAVY WILL HENCEFORTH BOARD AND INSPECT
ANY SUSPICIOUS VESSEL IN BAY OF BENGAL
NORTH OF 20TH MERIDIAN.
5. BRITISH NAVAL STATION, SCAPA FLOW,
REPORTED UNPRECEDENTED FREEZING OF
SEAWATER INSIDE THEIR BREAKWATER/HARBOR
FOR THREE WEEKS IN NOVEMBER/DECEMBER.
6. THIS REPORT IS CLASSIFIED IN
ACCORDANCE WITH EXECUTIVE ORDER 14708
AND INFORMATION IN THIS REPORT MAY NOT
BE RELEASED TO THIRD PARTIES WITHOUT
AUTHORITY OF NSC, WASH, DC.
XGDS-2
XXX

———

December - Boulder, Colorado

"Christ!"

Bart Jastre, like many of his English countrymen, had never been accused of being a man of few words. But now he found himself without words to express the shock and fear caused by the display before him. It was perhaps a sign of the times that a computerized audio-visual display could elicit such emotions, but it had. However, such displays had been endowed with the ability to portray the unseen, the previously unimaginable, and it was this previously invisible menace he now confronted.

Jastre sat at the conference table, as he had so many times before. To his left sat Ted Henke and Ken Brady. Conchita stood at the computer console which animated the wall displays Henke had montaged together.

"Could we see them all again. Same order as before, but slower?" Ken Brady requested.

"Sure thing. Just a sec," was Conchita's response.

Conchita typed a rapid string of keystrokes into the computer, waited a few seconds for the data to be accessed and then another string. After a moment's hesitation to check the screen Conchita paused and said, "Here goes," and hit the Enter key with a flourish.

On the wall screen before them the satellite images of the North Atlantic again appeared. To their trained eyes the rainbow bands of the computer enhanced infra-red imagery which whirled and spun across the ocean were implicitly clear in their meaning. The brighter oranges, reds and yellows represented the warm currents, the Gulf Stream and the currents at the Horse Latitudes, among others. And the blues, greens and purplish blacks showed the cold currents out of the arctic waters and the mid-sea doldrums between the clearly warmer eddies. At times the patterns of the bright colors shifted somewhat and whorls of color broke away and created spirals of color until they faded out, but the overall path was clear.

The white and blue labels at the bottom of the screen ticked off the year and date of the images. Each year took about twenty seconds and Henke had compiled ten years of satellite data which he had strung together. When the most recent year's images rolled around Jastre sat up in his seat. He was ready for it this time.

In the spring months it was clear that the eddies and whorls of color were numerous on the east side of the Gulf Stream, the huge orange yellow river running from Florida to Scandinavia. And by mid-summer the flow was clearly being interrupted and agitated in mid-ocean between Labrador and Ireland. The oranges and yellows turned into numerous red swirls and mixed with the blues and greens of farther north. And by the end of summer and into the autumn months a clearly new pattern had developed.

The main yellow stream which flowed past the east coast of the U.S. now went east off the coast of New England and Newfoundland, across the Atlantic and then turned south skirting the European mainland near France before joining the warmer waters to the south and hitting the westerly currents off North Africa. In place of the northerly leg of the Gulf Stream which had been clearly defined in previous years as extending as far north as Scandinavia and surrounding Iceland, there was now a well-defined counter-clockwise current in the North Atlantic. Arising in the black waters near Greenland, purple slowly growing into dark blue as it flowed south and met the Gulf Stream off of Newfoundland. It then warmed slightly as it interacted with the warmer water of the Gulf Stream achieving a medium blue color by the time it took a left turn to the west of Ireland and headed north. With a slow curve west in the waters around and north of Iceland the new loop, once again deep violet, finally returned to its source in the frigid waters south of Greenland.

After a moment of silence, Brady again spoke, "Give us a still shot of early December last year along with early December this year."

Conchita again typed and presently two images, now halved in size, appeared before them. The images differed radically. The broad golden band aimed at Norway the previous year was now a large oval in the Central Atlantic.

Jastre spoke, "Have we been able to determine a cause?"

Henke responded, "We haven't been able to analyze everything yet, we just now put all the pieces together to see this, but it appears to be a disruption in the interface zones between arctic and sub-tropic water and the varying salinities of the waters. We know that's what makes the deep ocean currents work and it appears to be what is at work here. If I were to hazard a guess it would be that an overburden of excess run-off from melting Greenland and north Canadian ice packs and the general warming of all the ocean thermals has overpowered the ability of the North Atlantic subduction zone current to suck the Gulf Stream in and the Gulf Stream has taken the line of least resistance by heading east toward central Europe and mimicking the Japan Current out in the Pacific in its clockwise oval flow."

Brady now turned to Henke, "I don't suppose we don't have any idea yet as to whether it is permanent."

"Nope. It might be that the cold currents circling into Greenland might have a chilling effect there to stop the melt and this thing could right itself in time, but I would not bet on it and the preliminary rerun through GAIA doesn't show it.

"The Gulf Stream really doesn't have that much effect on the parts of Greenland that are the major ice packs. The climate there is controlled by the prevailing arctic systems, out of the west, Canada, Alaska and Siberia. And the generalized global and arctic warming patterns probably won't be effected by this change in the Gulf Stream. It is the temperate areas formerly on the receiving end of the Stream that will be affected and are being affected. The northern half of Europe is in the worst cold snap in history."

"Well, gentlemen," Bart Jastre said, "If this continues for much longer, when I return to Merry Olde England, I may have to purchase an igloo to replace my cottage in Cambridge."

––––––––

December - Southern Connecticut

This time Molly drove the last stretch from Essex to the house. It was the only way she could enforce her command of "Don't open your eyes until I tell you." Sure, it was a bit silly, but Molly wanted everyone to take in her masterwork all at once and she wanted to see the reaction. Everyone, in this case, included both Sandy, on Christmas break from Columbia University, and Chad, having taken leave for the holidays from his flight training in Texas.

The masterwork Molly wanted to show them was the rebuilt family residence. She had spent the last five months coordinating between her insurance company and a contractor and architect. When her duties at the firm would allow she had flown up for the day to see the progress and give her input. Although Andy knew what she was doing, generally, he had been too busy to pay any attention to her project. Sandy had come up with Molly once to sort through the personal effects the wrecking crew had collected from the storm damage, but she had not been involved in the rebuilding. This would be the first view of the new house for the family.

The rutted farm path they had driven up in July was now a graded, gravel road. Andy, Chad and Sandy obeyed her wishes and covered their eyes for the last drive up the long hill to the house. Molly pulled up and stopped on the far edge of the paved courtyard. She told everyone to open their eyes.

The reaction from all three was the same, "Wow!" Andy added a "You did well," and reached over to give Molly a kiss.

Although the style was in general the same, the house, instead of a big old New England farmhouse, was now a stylish Cape Cod styled mansion. And in place of the old garage was a new garage that was really a garage instead of a barn, and it was styled to match the house, gables and all.

"And you mean to tell me that you paid for this with insurance money from the old house?" Andy asked.

"Well, I added a little capital, but not much. It is just that I am a good attorney and had my family covered by a replacement cost policy. The insurance company was a bit surprised to find out how much it costs to replace a two hundred year old house." She paused for effect. "Especially one owned by the senior U.S. Senator from the home state of the insurance company."

"Mom, you didn't!" chided Sandy.

"Naah, you know I wouldn't do anything unethical. Let's just say I didn't take any guff from the claims adjustor."

Molly laid her head back on Andy's collar, turning her forehead into his chin. She and he were laying back in the cushions of the big overstuffed love seat in front of the fire. This was one of the possible uses she had contemplated when she had selected the furniture.

There in the den, the sound of the television from the family room could barely be heard . It seemed that neither university student or rookie pilot had had much time for television recently and they were renewing an old acquaintance.

"Remember how many times you bugged them to turn off the TV and do something useful when they were kids. It seems that both our children have done pretty well in spite of the TV." Andy commented.

Molly smiled, "See. If I hadn't bugged them when I did, God knows what they might have turned out like."

"Uh huh." Andy gave her a quick hug.

"We haven't had much time to talk recently," Molly said.

"Obviously not, considering what you were able to do with the house without me knowing about it."

"You know I would have told you if you had wanted to know. You just were so preoccupied, I didn't bother. And besides, we all enjoyed the surprise."

"And it was quite a surprise. I couldn't have imagined anything so nice, especially considering what it looked when I last saw it."

"Well, it was just as I imagined it. I think a little girl, from the time she gets her first dollhouse, dreams of doing something like this. And here I had everything I could ever want to choose from, any style, any furniture. My choice."

"The career woman finally gets to play house."

She nudged him playfully with an elbow and then snuggled her back deeper into his side.

After several minutes of silence, watching the fire, she spoke, "So what is the big preoccupation you have. Still the environment."

"Yeh, as always. Last spring I had all the big dreams that at last this GAIA data would break things loose. But, after the first big rush, a few new regulations, and then the Miami riots, we're back to business as usual. And with the fear of starting another Miami fiasco, the President has put the lid on the most important information we have."

"What's that?"

Molly sensed Andy's hesitation, "If you can't say, don't worry. I'll understand," she reassured.

"No, much greater secrets than this regularly cross pillows in Washington. The President asked me not to make this public, so my kneejerk reaction was to clam up just now. The problem is that he is afraid the word that the global warming and ocean rising may be coming quicker than predicted will get out, and we don't have the full story on it yet."

"How soon?"

"Well, that's what we don't know. Even those who really believed in the theory always said it was perhaps one hundred or two hundred years away. Now it looks like maybe the radical who said the middle of the century, 2050, might be right, or even worse...."

"How worse?"

"Again, that's the mystery. The heart of the information the President is keeping quiet is that the Ross Ice Shelf is moving, slowly, but moving. But we just noticed it and it could be that it has been doing this for eons. But it doesn't look good."

"So he's keeping this quiet just because of Miami?"

"No, that's too simplistic. Actually, there are several good reasons to keep it under wraps. Economic, political and even military. The CIA has a theory that if a low-lying country knows it is going to be flooded it will invade a more elevated neighbor. It is a bit crazy, but in this world we live in, it probably has merit."

"And so, you know this is happening, or are afraid it is, and want to put programs in effect to stop it or slow it down, but you can't let anybody know why you want the program."

"Right, and several of the most important programs are political nightmares, unless we are able to scare the people, and the world, into taking the bitter pill. But if we tell everyone we think the ocean is really rising and we are wrong, we can really screw things up."

"Quite a dilemma you've got yourself there, Senator."

"Yup, quite a dilemma."

———

December - Ross Station, Antarctica

VIA NAVCOMMFAC - 2 PAGES XMITED
S E C R E T (XGDS-2)
FM: CDR, (ATTN:CAPT MARTIN), ROSS
STATION, ANTARCTICA
TO: GAIA OFFICE, NCAR, BOULDER, CO
INFO: SCRIPPS INST,(ATTN: DR. COLTON) SAN
DIEGO, CA
GAIA DISTRIBUTION LIST GOLF
USCGSTA,(ATTN: COHEN), NEW LONDON, CT
NSC, WHITE HOUSE, WASHINGTON, DC
SUBJ: STATUS REPORT - ROSS SHELF AREA -
DECEMBER 22ND
1. SUMMER SOLSTICE OCCURRED AT 0130Z
TODAY MARKING THE MIDDLE OF THE 24 HOUR
DAYLIGHT PERIOD. TEMPERATURES THUS FAR
HAVE AVERAGED SIX TO TEN DEGREES ABOVE
THE HISTORICAL AVERAGES, HOWEVER NOT
SIGNIFICANTLY ABOVE LAST YEAR'S RECORDS.
2. ALL SIX OUTLYING DATA ACQUISITION
STATIONS WERE OPERATIONAL THROUGH DEC
10TH WHEN STATION FOUR WENT OFF LINE.
INVESTIGATION REVEALED THAT THE WELL
CASING, EQUIPMENT AND CABLES WERE SEVERED
AT THE THIRTY FIVE HUNDRED FOOT LEVEL.
ACOUSTIC ANOMALIES IMMEDIATELY PRECEDING
CESSATION OF INSTRUMENT SIGNALS INDICATE
THAT EQUIPMENT WAS LOST DUE TO SHEARING
ACTION OF ICE/ROCK INTERFACE WHERE
SENSING EQUIPMENT FOR STATION FOUR WAS
LOCATED. RECOMMEND REPLACEMENT OF STATION
FOUR AS THIS IS THE STATION CLOSEST TO

OPEN OCEAN AT BAY OF WHALES.
3. AIR RECONNAISSANCE REVEALS SIGNIFICANT
SURFACE DEFORMATIONS OF ICE PACK TO THE
SOUTH OF ROOSEVELT ISLAND (160W-78S)
OVER THE PERIOD OF OCTOBER TO DECEMBER
CORRELATING TO USGS SEISMIC REPORTS.
RECOMMEND CONFIRMATION BY SATELLITE
IMAGERY.
4. THIS REPORT IS CLASSIFIED IN
ACCORDANCE WITH EXECUTIVE ORDER 14708
AND INFORMATION IN THIS REPORT MAY NOT
BE RELEASED TO THIRD PARTIES WITHOUT
AUTHORITY OF NSC, WASH, DC.
XGDS-2
XXX

———

January - Washington, D.C.

Senator Andy Knowles was on the hot seat. He had known it was a set-up when the Senate Majority Leader had called him, but Andy had come anyway. There was no other choice.

Now, he stood with his back to the others in the room, staring out the window onto the snow-covered Mall. His body language was not lost on anyone present.

Behind him, along with the Majority Leader, were the four ranking member from both parties on the Joint Conference Committee, the Speaker of the House and Rich Claypool, Andy's young compatriot from Connecticut, who was the working sponsor of Andy's bill in the House.

The other men all sat around the Majority Leader's long mahogany conference table. A Republican senator from the Conference Committee was at the far end and was speaking, "Its not like any of us disagree with the spirit of your position or the necessity of getting law like this on the books, sometime or in some form. It is just that we have to face political reality, as important as it is to clean up the air and stop the greenhouse gases, you just can't throw fifty thousand miners out of work along with the resultant repercussions in the industries that use the coal until you have an alternative."

Knowles turned, red-faced, at the senator, "Oh, cut the crap, I never thought I'd hear a good Republican like you spewing United Mine Workers propaganda. You know god-damn well that this bill doesn't put fifty thousand miners out of work. It is phased in over five years and just requires a reasonable shift toward alternatives to coal, or CO_2 scrubbers for the target industries."

"But, at the end of five years, whatever else, you will have all those miners either out of work or praying that Detroit and Pittsburgh have decided to buy the scrubbers," the Republican replied.

The West Virginia senator's input was "And, Senator Knowles, you have to remember that sixty percent of the electorate in my home state are either U.M.W. members or are related to one, and those mining folks all really do believe your bill will put them on the streets, jobless and homeless."

Knowles smirked, "And one hundred percent of my electorate rely on the environment for the air they breath, the food they eat and the life that they and their children want to live. When this fucking planet goes down the tubes what kind of political reality are you going to tell me about then?"

"Alright! I don't think this is going anywhere." The Speaker got up and walked to Knowles' end of the room. "Andy, all we are trying to tell you is that as much as anyone would like to help you get the Greenhouse Gas Bill passed and even though you may be able to put together a bare majority in the Senate, that there is no way this thing will pass the House. Sending the conference report as you proposed that it be drafted is pure futility. Rich?"

Rich Claypool cleared his throat and spoke, "Andy, I tried to tell you, but you didn't want to hear it. I have worked my heart out on this and I just can't muster the votes. They have a bi-partisan block that is more scared of unemployment than the environment."

Andy Knowles nodded and looked at the men gathered at the table, "So what are you presenting as an alternative?"

Everyone else waited for the Majority Leader to speak, it was clear that his state's vested interest and his position of power in the Senate would dictate the outcome of this compromise.

The Majority Leader locked eyes with Knowles for a few seconds before he spoke, simply and directly, "You can get a ten year phase-in for industries with over one hundred employees, with a provision for executive waiver for another five years in case of demonstrated extreme hardship."

Knowles sputtered and said, "You know that won't ..."

"Andy!" The Speaker cut him off. "Do you want half a loaf or none?"

Andy turned his gaze to the men who watched him. As he caught his eye, Rich Claypool raised his hands in a gesture of futility. Andy slowly shook his head. "Well, Gentlemen, I guess you know the answer. Once again we are going to give the most important issue we have ever faced a 'half a loaf'. We emasculated the clean water and air bills with regulatory exceptions for years. We laughed at do-gooders who cared about a silly little spotted owl. We looked the other way when Brazil, Indonesia and a dozen other countries denuded the rainforests that are the lungs of our planet. Our country was an obstruction at the Rio Conference instead of a leader in taking action on the environment. We smiled all the way to the bank while our multi-national corporations shipped our poisons and pollution to the Third World, as though as polluting the planet doesn't matter if we can't see it happen at home."

Andy paused to take a breath. "Now, of course, we will come to this great compromise that does mere lip-service to this, perhaps the greatest problem facing our environment; the greenhouse gases that are slowly baking our planet. We will put money and votes into political perspective and instead of retraining miners and retooling industries in ways that will help us stop the deadly greenhouse gases, we will pass the problem on to however has to make the hard choices ten, or maybe fifteen, years from now.

"But, Gentlemen, when you go home tonight after reaching this great compromise, you had better get down on yours knees and pray that God gives us that ten or fifteen years before our actions screw everything up once and for all. And you know what might be a good idea for each of you?" Andy stopped and walked back over to the window overlooking the Capitol Mall. "Before you leave tonight, take a walk down the Mall to the Jefferson Memorial and sit next to that beautiful Tidal Basin out front of the memorial. Do you know that the Tidal Basin doesn't empty out at low tide as much as it used to when it was built? I'm told that the reason the water is a bit higher is due in part to the burning of coal from your home state and the global warming it has caused so far," glancing at the Majority Leader, "and I'll bet that when your descendants have to take a row boat to get from the Capitol to what is still visible of the Jefferson

Memorial above the waters of the Potomac that they won't look too kindly on your great compromise here this afternoon."

There was silence as Andy Knowles walked out.

———

January - Bayonne, New Jersey

Captain Furosawa stood on the bridge wing of the Dekiru Maru. He could easily see the problem through his binoculars. He had read about this in a feature magazine back home in Japan, but he had never imagined he would get involved. Not like this.

It was easy to recognize her through the binoculars. The beautiful blonde in a red dress with the bullhorn. She looked just like her photo. He wished he could say he recognized her from her movies, but he could not. The American made-for-television movies in which she had made her name were not high on Captain Furosawa's list of entertainment. But, the circus on the Bayonne pier was definitely entertaining, even if it was interfering with his ship's business.

The crowd was certainly bigger than yesterday. The publicity from the earlier confrontation had brought on the curious to add to the serious protesters. And there were more police and the private security guards the Company's New York office had promised.

Right now below him the actress was speaking to the crowd. Furosawa knew English fairly well, but he could not understand what she was saying with the distance and the sound bouncing off the warehouse walls on the wharf. But, he could imagine what she was saying.

After the debacle between the stevedores and the protesters yesterday afternoon, Furosawa had met with the agent for the Company who had explained the situation before leaving to arrange for the security and countermeasures. The Company's offices in America were getting adept at handling the protests, not only at the shipping activities but also for the automobile and electronics companies who were likewise subsidiaries of the huge Japanese conglomerate.

Just as Furosawa had read in the magazine, the shipping agent had explained that the American activist groups had chosen the Company as its focus to protest the cutting of the rain forests. It seemed that someone had determined that the Company owned subsidiaries in Brazil who both cut timber in the rain forests for shipment and processing and then used the vacant land for beef production. This was nothing new to Furosawa, he had been shipping timber and frozen beef from Sao Paulo to Japan for years. In fact, this side trip to bring nitrate chemicals from New Jersey to Brazil was one of the few exceptions to the Brazil to Japan route.

However, this news of the Company's rain forest connection had spread like wildfire through the American environmental activist community. The Company was now being blamed for everything from the deforestation to the methane produced in the Brazilian cattle intestines contributing to the global warming.

Now the Company's business activities in America were being picketed regularly; its auto dealerships, its electronic stores, and most recently, someone had decided to target the maritime subsidiary which owned the Dekiru Maru, and the dozens of other freighters and tankers which were the prime carriers of the Company's products. And the blonde actress had made the fight with the Company her "cause."

Things were starting to happen on the pier. With the blonde's speech over the protesters had spread out to lay down in the tracks of the loading cranes, their tactic yesterday, too. The stevedores and longshoremen stilled lolled on the containers and forklifts, watching the show. They were obviously on Company time and were in no hurry for the spectacle to end and the work to begin.

Between the idled workers and the protesters the police and security men fanned out. One of the policemen was using his own bullhorn, but it was barely more audible than the one the actress had used. Furosawa was able to pick out something about "this pier is private property" but nothing more.

At the head of the pier Furosawa could now see a line of blue police vans and buses drive up and park. The first van continued on and made a u-turn to back up near the protesters. Two officers got out and opened the rear door.

The officer with the bullhorn said something else. It was obviously a last warning, since the lines of police and security at that point waded into the protesters and started to try and extract them from their positions.

———

Alexandra "Sandy" Knowles sat shivering on the asphalt. She had worn the long pants and warm clothes the protest organizers had recommended, but it was not much help. It could get mighty cold on the New Jersey waterfront in January. And she had inherited her father's long legs and sitting indian-style on pavement was not very comfortable.

The bus trip down from the Columbia University campus had been interesting. From the attitude of the people on the bus, you would have thought they were children going to summer camp rather than college students going to protest the poisoning of the earth. A couple of people had brought guitars along and had gotten everyone singing, mostly old folk songs and the protest group standards of "We Shall Overcome" and "Blowing in the Wind", as well as a few of the newer nature-oriented protest songs that nobody seemed to know the words to. It did serve to keep everyone's mind off of what they would face in Bayonne.

Now, as Sandy sat on the pier listening to this woman speak, she did not know whether she resented her glamorous presence in their midst or if the media attention brought on by the famous actress was important to what they were doing. The woman was certainly delivering her lines well in the speech she was giving, but Sandy had the nagging suspicion that this was exactly what they were to the actress, lines on a script. Sandy wondered how the others felt about this, they did not appear to care. Now was not the time or place to discuss it. She would have to ask the others, her friends, when they got a chance after this. Perhaps in jail, where they expected, or perhaps hoped, to be after the protest.

Sandy wondered whether her skepticism, which apparently differed from the other protesters, was perhaps a result of the fact that fame and publicity were not as foreign for her as it was for the others. She had spent much of her teen years in the limelight of her father's presidential bid and as far back as she could remember he had been in national politics. Her father's campaign had drawn dozens of celebrities just like this actress to it. At Senator Knowles' side Sandy had personally met Mother Theresa, Nelson Mandela and the leaders of a dozen nations. This soap opera queen did not impress Alexandra Knowles.

But now the soap opera queen had finished and the murmur of expectation ran through the crowd. Sandy and her college friends moved into their planned places on the railroad tracks used by the overhead crane. Assignments had been given out on the bus.

When they were in place someone nudged her with an elbow and said, "Let's lock elbows." And so they did, friends and total strangers, protesters all, locking arms to hinder the police who now started to wade into the crowd. When they threw the first person into the paddy wagon everyone knew they would all be arrested. It did not matter. They were prepared for that.

What would her parents think? Senator and Mrs. Knowles.

Her mother had told her of the anti-war and anti-nuke protests that both her mother and father had participated in at Yale. So, they could not really fault her tactics. But Sandy doubted that the gentile protest rallies her parents had probably staged at Yale University were anything like the rowdy free-for-all on longshoremen's turf she was now engaged in.

And her father was probably the foremost advocate of environmental action in the U.S. Government. He could not help but believe in her cause, even though she had not had occasion to talk to him about it. She had listened to his preaching on the environment for years now. No, her parents would not have any trouble with what she was doing. She was her parents' daughter and she would do what she must do.

All around her chaos reigned. There were many more police than in the video tapes they had seen from yesterday. As the police and rent-a-cops pushed further into the crowd someone tried to start up "We Shall Overcome" but it never really got started.

Finally the police reached Sandy's group on the railroad tracks. By now the police had ceased to ask each protester to move peacefully. No one had moved so far, so they knew they were wasting their breath.

Sandy and her neighbors locked their arms until the police, with an expert twist of a billy club, ripped them apart. When she felt herself being lifted up Sandy went limp as instructed, to make it harder for the police to carry her. Two rent-a-cops carried her about a hundred yards and dumped her unceremoniously on the floor of a dark blue jail bus where they were taking the women protesters.

As Sandy looked up from the floor she saw, towering above her, a huge black woman in a jail matron's uniform. The woman carried a shiny black riot baton.

"Get your ass in the back of the bus and don't give me no crap!"

Any thoughts of continuing her sit-in on the floor of the bus vanished at the thought of what this behemoth of a woman might do with that baton. Sandy, with all the dignity she could muster, got up off the floor and took her place in the back of the jail bus.

For the next hour Furosawa watched as the police struggled to clear the pier. They carried limp bodies and dumped them in bus and paddy wagon. An occasional scuffle or punching match broke out, but nothing serious. Toward the end Furosawa could even make out the red dressed torso of the actress being carried to a waiting police car. She had indeed been true to the cause she had rallied her followers to.

Furosawa marveled at the futility of it all. That these stalwart people would risk arrest and imprisonment for the few hours of delay their actions would cause his ship. These headstrong Americans; you had to love them. Certainly they did not think that something like this would dissuade the Company from carrying out its business.

Furosawa had been with the Company ever since he had gotten his Mate's certificate, nearly thirty years before. He knew the Company. The Company and most of its employees, at least the Japanese ones, were devoted to the success of the Company and its profit. The Company had a long history and had overcome much to get were it was today. The early Company, nearly a century before, had built the battle cruisers that had beaten Russia at Port Arthur. It had built Japan's first delivery truck under a license from Henry Ford himself. The company had built many of the planes that had fought in World War II. And after the war it had been one of the first to pull itself out of the ashes and send shiploads of merchandise to sell to its conqueror. Now the industrial base of the Company surpassed many large nations all by itself. Did these Americans think that the Company which had overcome the stigma of Pearl Harbor to achieve enormous profitability in the American marketplace would let this pretty little blonde in her red dress force them out of a lucrative business venture in Brazil? Hardly.

The pier was clear now and the workers were starting to load the ship. Pallet after pallet of the chemical containers with their warning labels were hoisted aboard. The Company needed these nitrates to grow cattle fodder down in the rain forest and Furosawa would get them there.

Furosawa knew that the Company would continue its business in Brazil so long as it was profitable. It had an enormous investment in Brazil and in the worldwide infrastructure to make use of the resources found there. The Brazilians were happy and the Company was profitable. The business would go on no matter what these American protesters did.

———

January - New London, Connecticut

VIA NAVCOMMFAC - 2 PAGES XMITED
S E C R E T (XGDS-2)
FM: CMDR COHEN, USCG, NEW LONDON, CT
TO: GAIA OFFICE, NCAR, BOULDER, CO
INFO: SCRIPPS INST,(ATTN: DR. COLTON) SAN
DIEGO, CA
GAIA DISTRIBUTION LIST ECHO
NSC, WHITE HOUSE, WASHINGTON, DC
SUBJ: SITREP OF SIGNIFICANT MARITIME
INFORMATION - DECEMBER
1. BRITISH AUTHORITIES ISSUED NOTICE TO
MARINERS REGARDING HAZARDOUS SEA ICE IN
SEVERAL LEEWARD AREAS OF IRISH SEA AND
NORTH SEA.
2. IRISH MARITIME OFFICE REPORTED DELAYS
IN PORT OPERATIONS IN BOTH DUBLIN HARBOR
AND SHANNON RIVER AREAS DUE TO FROZEN
HARBOR APPROACHES.
3. JAPANESE DEFENSE FORCE FRIGATE
REPORTED THAT IT HAD BOARDED AND SEIZED
A FRENCH TRAWLER REGISTERED TO AN
ENVIRONMENTAL ACTION GROUP AFTER IT
RAMMED A JAPANESE FACTORY SHIP FISHING
USING GILL NETS IN EASTERN PACIFIC AREA
ABOUT EIGHT HUNDRED MILES OFF CALIFORNIA
COAST. FACTORY SHIP WENT INTO PORT AT
SAN PEDRO, CA WITH AID OF SEA-GOING
TUG AFTER SIGNIFICANT LOSS OF FUEL OIL
FROM COLLISION DAMAGE. OTHER JAPANESE
VESSELS LEFT AREA, HOWEVER THIS APPEARS
TO BE DUE TO LACK OF FISH CATCH IN AREA

RATHER THAN ACTION BY PROTEST GROUP.
BUREAU OF FISHERIES REPORTED CATCH OF ALL
COMMERCIALLY SIGNIFICANT FISH CATEGORIES
IN EASTERN PACIFIC WERE DOWN FROM
SEASONAL NORMS.
4. RUSSIAN OCEANOGRAPHIC JOURNAL,
OKIANSKIJ ZHURNAL, REPORTS RESULTS OF A
STUDY IN WHICH THE CARBON DIOXIDE CONTENT
OF SEA WATER AND OCEAN TEMPERATURE
WERE STUDIED. RESULTS INDICATE DIRECT
CORRELATION BETWEEN ATMOSPHERIC AND
OCEANIC CO2, AS WELL AS COROLLARY WITH
HEAT RETENTION AND THERMAL EXPANSION AS
OCEAN CO2 LEVEL INCREASED. RECOMMEND
REVIEW BY GAIA STAFF.
5. RESEARCH THESIS BY USCGA SENIOR
MIDSHIPMAN INDICATES THAT WATER POLLUTED
BY EITHER PETROLEUM HYDROCARBONS OR
SEWAGE PARTICULATES HAS THERMAL RETENTION
OF UP TO FORTY PERCENT GREATER THAN NON-
POLLUTED FRESH OR SALT WATER. THESIS TO
BE FORWARDED UNDER SEPARATE COVER TO
GAIA, BOULDER.
6. THIS REPORT IS CLASSIFIED IN
ACCORDANCE WITH EXECUTIVE ORDER 14708
AND INFORMATION IN THIS REPORT MAY NOT
BE RELEASED TO THIRD PARTIES WITHOUT
AUTHORITY OF NSC, WASH, DC.
XGDS-2
XXX

February - Ross Ice Shelf, Antarctica

Captain Daniel Martin paced anxiously behind the radio operator. Things were not going well. Although the SATCOM Network theoretically had world-wide coverage, there were times when out of the way places could not come up on the network easily. And he was in about as out of the way place as could be. The ultra-high technology multichannel satellite system was sometimes as cantankerous as the old HF radio system it replaced.

The radioman continued, "Golden Eagle, this is White Maiden, over."

The radioman had repeated the call-up another three times when the speaker crackled to life.

"White Maiden, this is Opera Box. Over."

"Opera Box, this is White Maiden," the radioman responded as he reached for his JANAP book to find out who 'Opera Box' was. Martin looked over his shoulder and saw him point is fingernail at 'DOD-Central Command HQ, McDill AFB, Florida.' The radio crackled back to life.

"Roger, White Maiden, this is the watch officer at Opera Box. We have been hearing your calls to Golden Eagle. Our references show that Golden Eagle is not active on this circuit except during business hours on weekdays Eastern Standard Time, at all other times Full House acts as a patch for them. Over."

"Roger that, Thanks." The watch officer at McDill Air Force Base in Florida had broken circuit discipline to steer them in the right direction. The skeleton crew at Ross only had the most basic of communications gear and did not have the right information to handle all the possible exigencies of global communications. The stuffed shirts at Full House were probably sitting there having a good time at their expense, not answering because the call-up was not directly to them. This was not a good time for gamesmanship or overly rigid discipline. And it was not like there was any radioman in the military that did not know who Golden Eagle was. Neither his radioman nor Martin needed to look up who Full House was, either. It was the National Military Command Center, NMCC, at the Pentagon.

The radioman started the proper call-up, but Captain Martin took the microphone from him. Before he could speak another tremor shook the communications shack. The radioman steadied in his rolling chair by grabbing the table. The base commander stood balancing with feet spread like he was on the deck of a rolling ship at sea. Martin waited until the slight vertigo from the unexpected motion had ceased.

Captain Martin depressed the hookswitch and spoke, "Full House, this is White Maiden, Over."

"Roger, White Maiden, this is Full House, Over." The churlish lilt to the voice was unmistakable, even over seven thousand miles and having bounced off a satellite.

"Full House, have you been hearing our calls for Golden Eagle for the last ten minutes? Over."

"We have, but SOP authorizes us to answer only calls directed to us, even if we are patching for other stations. We cover a lot of bases at this time of night. It is the caller's job to notify us of a patch. Over."

"When you are answering calls for the Chief, it isn't smart to play games. We have Flash traffic, code name Pinnacle, for Golden Eagle. Repeat, Flash Pinnacle traffic. Quit being a horse's ass sitting on the SOP and patch us through to Golden Eagle and put the Duty OPS Officer NMCC on the line also. Now! Over."

Martin listened to a series of clicks and beeps as the chastised operator at NMCC patched the call through to both the Duty Officer at the White House and the general officer on night duty at the Pentagon. The wait was interrupted by another sickening lurch of the ground, or rather ice, beneath Ross Station.

———

February - Washington, D.C.

Andy Knowles had never before travelled Massachusetts Avenue without traffic. But, he had rarely driven in from Bethesda to central D.C. at two in the morning in a snowstorm either.

Molly had taken the call not more than twenty minutes before. She had trouble waking him up. Andy had put in a late night on the Hill and was still in a deep sleep when the simple message got through to him.

"Senator, this is the staff duty officer at the White House Situation Room. The time is now 0210 Monday morning, are you awake, sir?"

"Yeh. yeh. I'm awake."

"Thank you, senator. The president has called a meeting in the Situation Room at 0300 hours and requests your presence. The meeting is in fifty minutes. Will you require transportation?"

"No, no. I can make it in."

"Senator, our procedure is to ask you to repeat the message back so we can be sure you understand and are awake. Could you do that for me, Sir?"

Andy was fully awake by now. "OK, there is a meeting in the Situation Room at 3 o'clock and I'm to be there. Right?"

"Thank you, Senator. If you would like to confirm this message you can call the White House switch board at 456-...."

Andy interrupted, "That won't be necessary."

"Thank you Senator."

At Andy's recitation of the message Molly sat up all the way in bed and flipped on the light on the bedstand. When he hung up, she spoke, "What's this all about?"

Andy turned back the covers and headed to the bathroom, "I don't know, but it can't be good news."

Now, he drove through Georgetown in silence. The only thing moving were the D.C. city snow plows and sand trucks.

D.C. was notorious for letting just a little snow snarl traffic and this was more than just a little snow.

Andy finally circled the White House. Having been here many times over the past year he turned into the executive parking garage across the street from the White House. The little known tunnel across to the White House would get him across the street without going into the snow.

The Situation Room was, as usual, humming with activity, but it was not full. Obviously, the late night call had not been fully successful, or perhaps everyone had not been called or could not get in due to the storm.

Andy, himself, arrived just in the nick of time. The Vice-President was there and Andy had just shook hands with her and was starting to speak when Admiral Kirksey's deputy asked everyone to take their seats. Andy started to go up to the upper row when the Vice-President motioned him to the seat next to her. "The President has already been briefed and not many of the Cabinet are coming in at this hour. Have a seat in front."

Andy nodded and sat next to the Vice-President and the Energy Secretary took his seat next to Andy. No one, of course, sat on the other side of the Vice-President in the chair with the embossed seal.

A member of the National Security Council staff, one of the civilians, took the podium. He shuffled a stack of papers and began, "Madame Vice-President, members of the Cabinet, ladies and gentlemen. We have already briefed the President on these matters and he has requested that you all be informed as soon as possible. That is why we had to call you out on this miserable night.

"At 0539 hours Zulu, that is about five hours ago, the US Navy station on the Ross Ice Shelf reported severe tremors and possible movement of the Ice Shelf. That was followed up approximately a half hour later by reports from McMurdo Station, which is very near the edge of the Ross Shelf, of seismic activity.

"Immediately thereafter the commander of the Ross Station reported that the satellite positioning system in one of his Snow-Cat vehicles showed that the station was approximately seven miles from the previously fixed location. The positioning system is accurate to within a matter of meters and virtually foolproof.

"An early morning, Antarctic time, air reconnaissance of the region from McMurdo Station revealed that there were massive fissures of the edge of the Ross Shelf area nearest McMurdo and numerous instances of major calving, that is a breakup of the edge of the ice pack, on the northernmost, seaward, edge of the Shelf. The National Imaging Center has recently confirmed by satellite that there are fissures several miles wide along the Queen Maud Mountains and the Shackleton Coast, that is the southern edge of the Shelf, where it attached to the Antarctic continent proper. We are reproducing some charts and photos of this which will be available to you in a few moments.

"What this means is that the Ross Ice Shelf has done the unthinkable. It has slid several, at least ten, miles into the ocean over the last several hours. Satellite imaging indicates that the width of the movement although somewhat fragmented, covers a front of almost 230 miles. You can do the math yourself. If you displace a ten by two hundred thirty mile area, that is, twenty three hundred square miles of ocean, over a short period of time, you will have a significant oceanic disturbance. I think all of you have been briefed on the tsunami effect.

"We have every reason to believe that a tsunami is in progress in the Pacific at this time. One of the papers you are being given," he motioned to a staff member to commence handing out the papers, "is a listing of the probable arrival times of the tsunami at different places in the Pacific."

The briefer paused for emphasis, "Lastly, I need to tell you that this is precisely the kind of occurrence which can be the onset of the kind of catastrophic sea level rise that was discussed last year, assuming this continues and we see nothing to indicate it won't. The Queen Maud Mountains are nearly four hundred miles from leading edge of the Ross Ice Shelf. I am told that a four hundred by two hundred thirty mile chunk of the Ross Ice Shelf has a sea level equivalence of about five to seven meters, maybe more with the secondary breakups it can cause. That means that if this slide continues it could bring on a twenty foot or more increase in sea level worldwide.

"Although most of the personnel are in the process of evacuating Ross Station, the commander and three other crewmembers are staying on, for the time being, to ride this out. We spoke with them shortly before this meeting, that's six hours after the first report of movement, and they indicate that the tremors and movement are continuing."

The briefer spoke quickly with another NSC aide and then continued, "What we are going to do now is give you some time to peruse the material we have given you and get a cup of coffee. Let's say ten minutes. Then we will be continuing with the briefing including the operations which are being put in place to deal with this. Thank you. Coffee and I believe some donuts are going to be available in the anteroom."

Andy looked next to him to the Vice-President. She was sitting forward with her elbows on her knees, her hands covering her face. Andy understood the feeling.

———

February - San Diego, California

The quarterdeck crew of the USS Ronald Reagan was having atypical four hour watch. Sunday evenings were always slow, especially when the crew was on standdown. The huge aircraft carrier had just come back to San Diego from a deployment to the Far East and Indian Ocean. The ship was undergoing what was known as a maintenance availability. Most the ship's engineering equipment was off-line and the ship was relying on services -- water, power, heat -- from the pier at North Island Naval Air Station. Much of the ship's electronic and mechanical equipment was being worked on and in some state of disassembly by twenty-odd private contractors. The crew, both officers and men, always took advantage of these periods to take leave. It was a rare occasion when the crew, let alone senior officers and chief petty officers, could count on a few weeks when they knew they would not be needed.

Lieutenant (j.g.) Clifton Tolley was the Officer of the Deck, the O.O.D. Together with the Petty Officer of the Watch, the P.O.O.W. or better known as "the Pooh", and a couple of enlisted men, the O.O.D. spent a four hour watch on the quarterdeck, which on an aircraft carrier was a whole in the side of the ship just forward of an elevator, served by a long stairway to the pier.

Quarterdeck watch was not bad duty, just boring. The last watch, four to eight, had reported nothing in the quarterdeck log more exciting than the return of two drunk sailors by the Shore Patrol from the Coronado Amphibious Base Enlisted Men's Club. Tolley expected a quiet watch before being relieved at quarter to midnight.

The quarterdeck crew had two radio speakers to listen to. One was the Harbor Common circuit which was the general use radio band for harbor traffic. It was turned down to next to nothing in volume since the moored Ronald Reagan had no business with the harbor traffic. The other radio speaker was SOPA San Diego. SOPA, or Senior Officer Present Afloat was the Navy radio circuit that linked all of the ships to the Admiral's headquarters at COMNAVSURFPAC, SOPA. Ronald Reagan technically did not belong to SURFPAC, it was an AIRPAC asset, but all Navy ships in San Diego were required to listen to the circuit. Tolley had never heard anything come over the SOPA net except for the hourly drills in which three ships were called and had to answer by landline telephone within a certain time or the ship's captain received a nasty note explaining the discrepancy along with the exact time so that the offending O.O.D. could be chastened for his inefficiency.

The first hours of his watch passed without incident. Several crew members and contractors passed through the quarterdeck, coming or going, on their evenings business. Various duty stations reported their routine reports to the quarterdeck and the Pooh dutifully logged them in the logbook.

A few minutes before eleven o'clock, the SOPA radio net crackled to life. Both Tolley and the Pooh walked over and the Pooh turned up the volume.

"All ships present San Diego, all ships present San Diego, This is COMNAVSURFPAC. We have Flash traffic. This is not a drill. Repeat, this is not a drill. Standby to copy." A pause. The Pooh grabbed a pen and paper.

"A Tsunami Warning has been issued for all Pacific naval installations. All ships, units and installations are ordered to immediately institute Phase One of Operation Order Tango Zulu, repeat institute Phase One of Operation Order Tango Zulu. All ships and independent units are directed to respond by landline and to come up on SOPA Tactical net, frequency one two niner point six five. I repeat..."

The radio repeated the announcement while Tolley ruffled through the accumulated junk in the quarterdeck podium to find the O.O.D.'s handbook.

Tolley had not had need to look in the O.O.D.'s Handbook since he had finished his watch qualifications. The handbook had all of the checklists, emergency information, important phone numbers, etc., that might be needed by the quarterdeck or the O.O.D. It also had all of the Op Orders for emergency sorties, nuclear attack, civil disturbance, and the like.

The quarterdeck crew and a few other crew members who happened by during the radio broadcast clustered around Tolley as he thumbed the notebook tabs to OP ORDERS and then found Tango Zulu.

The majority of the information in the handbook was well worn, dog-eared and finger-smudged from months or years of use, but OP ORDER Tango Zulu's sheets were sealed in a plastic document protector and unused. It had obviously been a recent addition to the handbook.

Lieutenant (j.g.) Tolley read for all to hear, "Executive Summary." He cleared his throat. "Operation Order Tango Zulu will be executed in response to the imminent danger of tsunami and coastal inundation due to breakup of large masses of polar ice. The Operation, in three phases, will involve emergency sortie of all ships to areas protected from tsunami effects and evacuation of endangered coastal facilities and personnel to safe areas. Phase One will involve preparation of all ships to get underway within the shortest time possible and emergency notification of all facilities and key personnel, Phase Two will involve sortie and facility evacuation and as resources are available, evacuation of dependents, Phase Three will constitute reestablishment of facilities and relocating units and personnel as necessary in post-inundation situation. Phase One will be conducted in a confidential manner without public notification. it is assumed that Phases Two and Three will be conducted concurrent with public notification.

"Op Order Tango Zulu was promulgated in response to Executive Order 14746 and Defense Operation Order 49702, which call for all government agencies and military units to prepare contingency plans to respond to an imminent threat of tsunami and coastal flooding from a massive displacement of ocean water in the event of breakup of polar icecaps, primarily Antarctic. These governing instructions require all units and facilities to prepare for a worst case scenario of a ninety foot tsunami and up to thirty feet of sea-level rise on the west coast of the United States and a thirty foot sea-level rise on the east and Gulf coasts."

As Tolley continued to read, the Pooh started to call the senior officers, both the few still on board and those at home. Next, the watch section of the crew on board would have to do whatever was possible to patch the partially disassembled ship together and run up the reactors to get underway. It would be a long night.

———

February - Newport Beach, California

Ian Petersen had been lazy on Saturday, which always meant extra work on Monday. He usually did the daily bank deposit for the restaurant himself, so he now had three days receipts on his desk to tally and run down the block to the bank. He had already finished checking the first of the month deposit of rent checks from the apartments.

It had been a good weekend for the restaurant, especially in bar receipts. The total receipts had been nearly thirty thousand, equally distributed between cash and credit card. His secretary and bookkeeper had finished the Visa and Mastercard packet and the American Express packet, and had the express pouches ready to send these credit slip accountings to Nevada and Denver, respectively. Ian still checked each before sending them off each day. But, he did the cash receipts himself. He saw no reason to tempt a young employee with fifteen thousand cash. Pilferage in the kitchen and short-tabbing by the cocktail waitresses was bad enough. There was no use asking for trouble.

He was finishing the last cash count when he heard tromping of many feet up the hallway to his office. Instinctively fearing the robberies that had hit several South Coast eateries in the past few months, Ian pulled the second drawer halfway open. The 9mm automatic was there and ready.

The tromping ended with his office door swinging open and his entire early morning staff running into the office. All three morning waitresses, the cook, table boy and his secretary/cashier/hostess, Delores, jostled into the room and went directly to the television on the opposite wall. Delores grabbed the TV remote from the corner of his desk and flicked the set on.

"What the hell is going on?" Ian asked as he rose from his desk.

One of the waitresses answered, she was out of breath from excitement and the run from the kitchen. "We didn't get the morning linen shipment, so we called the distributor, they said they'd heard the whole California coast is being evacuated."

Delores punched the controller to switch from HBO to channel 10. The pretty news anchor was clutching a stack of papers and looking somber.

"Let me repeat that list of areas affected. All coastal areas of Santa Barbara and Ventura counties. In Los Angeles county, all of Malibu, Pacific Palisades and Santa Monica, all of the Los Angeles south of the Santa Monica Mountains between the 405 Freeway and the ocean, all of the beach cities between Santa Monica through Long Beach and other communities south of the 91 Freeway, all of Orange County between I-5 and the ocean and all other coastal communities in the TV 10 viewing area southward in Orange and San Diego counties.

"Repeating once again, the Federal Emergency Management Agency and the California Office of Emergency Services have declared an emergency and ordered an evacuation of the areas we just mentioned due to a Tsunami or tidal wave warning. The tsunami which has already hit New Zealand, many pacific islands and the coast of South America is expected to hit sometime this afternoon. Everyone in the effected areas is ordered to evacuate immediately to unaffected areas.

"Emergency shelters are being set up for evacuees. Recommended evacuation destinations are the San Fernando Valley, the San Gabriel Valley and Riverside and San Bernardino Counties." The newslady took a moment to breath and Ian's employees all started to talk at once.

"I gotta go get my kids from school."

"You think the buses are running?"

"Christ, what do you think caused that?"

In the clamor, no one could hear anyone, and Ian could not hear what else the TV was saying. "Hey, hey, hey! Everyone calm down." His staff turned to him and Ian continued, "We have to get things closed up and get out of here. Who does not have a car and needs a ride? Besides Rafael."

"Yeh, me."

"And me, too. But I have to get my kid from school in Costa Mesa."

"OK. OK. Rafael, you have the keys to the delivery van?"

"Si."

"All right. You take it and give these gals a lift. And pick up Debbie's kid from school. Then get up to high ground. OK?"

The two waitresses and Rafael started to turn, but Ian had another thought. "Wait," he said reaching for the deposit bag.

"I know payday is Wednesday. And if this is as bad as they say, you may need some cash."

Ian counted five hundred dollars out for each of the six employees from the Saturday deposit bag. As each employee took the money, Ian either was hugged or shook hands.

He followed the employees out to the front. Raphael and the two women headed back to the kitchen to leave in the van and the others started out the front door.

"Everybody has my cellular phone number. Call me and let me know how you are. Get back in touch with me as soon as this is over. OK? Rafael, will you lock the back on the way out and just keep the van until we can get back together."

"Si, Mr. Petersen."

In just a moment Ian was alone in his empty restaurant. He heard Rafael slam and lock the back door. Ian turned toward his office, but instead went to the cash register and took the cash out.

As Ian was emptying the money into the cash pouch, two men in business suits came into the restaurant.

"I'm sorry. We're closed. Haven't you heard about the evacuation?"

"What are you talking about?" one of them asked.

Ian gave a quick explanation and the two men left in a hurry. Ian went back to his office and quickly put the cash and credit card packets in his gym bag. The television was still on, a reporter was speaking.

"Thank you, Andrea. I'm here at the National Guard Armory in Los Alamitos where I've come across some interesting information. One of the Guardsmen here, a military policeman, said that his unit received their initial call-up yesterday and they have been preparing for this evacuation since last night. Also, it appears that the Emergency Service authorities already have plans in place for this evacuation and this is perhaps not a surprise to them. I'll have more on that story when we can. This is Brad Durrant, TV 10 from the Los Alamitos Armory, back to you Andrea."

Ian clicked the television off, so he could think. Did he have everything he needed from the office? Should he make the deposit at the bank before he left? No, if they were having a tidal wave he did not want the deposit made in a bank two blocks off the beach.

Damn! Cyndi! He had not called Cyndi yet.

Ian already had his cellular phone in its carrying case so he lifted the phone from its cradle. He pushed the button to get a circuit, but was rewarded with the annoying staccato beeps that meant all the cellular circuits were busy. Everyone else was in their car calling. He reached for the office phone on his desk and was thankful he had decided to call before he left. He dialed.

"Petersen residence, this is Cyn..."

"Cyndi, have you heard?"

"Heard what?"

"The evacuation. They're evacuating all along the coast. Some kind of tidal wave. Coming this afternoon."

"Jeez'... What are we gonna do? What about the kids?"

"I'll be right there. You stay there and I'll get Amy first at her school. And on the way out we'll go up Irvine Avenue and get Crystal from the junior high."

"Ok, hurry. Anything I should do?"

"Yeh, load the Volvo up with clothes, food and, yeh, all the earthquake preparedness stuff, you know the olive drab boxes in the closet. I'll be there in five minutes."

"Ok, hon, I'll be ready."

Ian grabbed the portable phone, his briefcase and the gym bag. He thought for a moment and reached into the still open drawer and took the 9mm automatic and the box of shells.

At the front door, as he sat down his gear to lock the door, he noticed the deli case. It had already been stocked with the ready-made sandwiches for the lunch crowd. He grabbed a cardboard box and filled it with the sandwiches and salad containers. Then he gathered everything up and locked the door to his restaurant, perhaps for the last time.

The trip home took a lot longer than five minutes. Balboa Boulevard was moving at a snail's pace. Luckily when he got to Amy's school the teachers had the children all lined up out front, as though they were waiting for buses. When traffic stopped he was able to hop out and shout to Amy's teacher, who brought Amy over. As they approached, Ian got back in the car and opened the door for Amy. She had been crying. He gave her a quick hug, but traffic had started again and his delay brought a chorus of honks.

"Come on, Amy. What's the matter?" Ian reached over and wiped a tear off Amy's cheek.

"They told us about the tsunami and said we might have to leave without our parents. Where's Mommy?"

"She's at home getting ready for us. I just talked to her on the phone."

"And Crystal?"

"We'll get her from school we when leave."

"What are we gonna do?"

Ian smiled to himself. Amy's 'What are we gonna do?' sounded just like her mother's, pure southern California accent. And the Californians thought his Canadian accent peculiar.

"We'll just get up away from the ocean for now and then we'll see what happens."

"OK." Amy took a deep breath and settled back into the plush seat next to her stepfather.

Amy was out of the car running to Cyndi, before Ian had the car stopped. Cyndi had the cargo door to the Volvo station wagon open and was stuffing a bundle in. Mother and daughter hugged, then Ian got his hug from Cyndi.

"I'm all set," Cyndi said. "I got all the emergency supplies, extra clothes, food, the works. I even got the camping stuff you took to Canada last year. You took a lot longer than five minutes."

"The traffic was awful. We ought to get going, if we are going to get Crystal. It seems they're busing the students. Probably a good idea, lots of parents couldn't get there in time."

"I already locked up upstairs."

"Let's see. Is there anything else we need?" Ian looked at Cyndi. They both looked up toward their apartment and back. From her look, Ian realized that Cyndi understood they might never see any of their things again.

"Important papers? Valuables?" he asked.

"Got'em"

Ian had to move out of the way of one of his tenants as they left and waved to him.

"How about water? First aid?"

"It's all in the earthquake preparedness packs. Ian, trust me, I packed everything."

Ian nodded, she was right, he should trust her. It was just her beautiful bright eyed innocence sometimes covered the fact she was a sharp lady, in her own right. He smiled, kissed her and opened the door for her and Amy to get in. He motioned for her to follow his Mercedes and Ian had just started to walk from the Volvo when Amy gave a blood curdling scream.

"Tinker!"

Cyndi spun in the seat, and Ian stopped short and turned to Cyndi.

Cyndi put out her hand to touch Amy's hair, "Relax, sweetheart, Tinker is right under Mommy's seat. You know how she is in a car. She thought I was taking her to the vet." Amy smiled and set to work coaxing the cat from under the seat.

Ian smiled over to Cyndi. She had remembered everything.

Traffic was even worse on Newport Boulevard. A police officer had his car blocking off the off ramp to Pacific Coast Highway. It was just as well, when they went over the overpass Ian could see that PCH was bumper to bumper and dead stopped.

Traffic was not as bad as they approached the junior high. When they turned onto Irvine Avenue, he found out why. A tarpaulin was hanging across the main entrance to the school. In hastily painted letters it read, "Students are at Tustin Civic Center."

As They say Ian pull back into traffic, Amy asked her mother, "Where's Crystal?"

"They already took her across town where she'll be safe. We'll go get her there."

As they headed up Irvine Avenue toward Tustin, Ian realized he had not been listening to the radio, which he turned on. The radio reported numerous cases of looting in evacuated areas, huge traffic pileups and even a report of a bus being hijacked by a gang in Compton.

Traffic was very heavy between Newport and Tustin, but the wide boulevards of the well planned communities of Orange County carried the traffic well. It did not take long to get to the civic center in Tustin. Cyndi hopped out to look for Crystal while Ian and Amy waited in the cars.

It was while Cyndi was looking for Crystal that the radio broadcast the first report from Hawaii. The tsunami had hit an hour after dawn Hawaiian time. With the tidal wave coming from the south, Honolulu had been hit squarely. The radio reported that everything on the coast from Waikiki through the main airport and Hickham Air Force Base on over to Pearl Harbor had been devastated. The radio also said the Navy had had time to sail the navy ships from Pearl Harbor around Oahu to Kaneohe Bay where they were safe.

Cyndi returned with Crystal. Crystal got in with Ian. Crystal gave Ian a one arm hug from the passenger seat. With the family together again, Ian wasted no time getting to the 55 Freeway north and when there, coming to a complete stop. Several million other people were also heading for safe, high ground, and unlike the city streets, the freeways of Orange County are notoriously prone to stoppages.

However, it was only a few moments later that a highway patrol officer motioned Ian's lane to the left across the median and up the south bound lane. They had closed down the freeway from the other end. It was a bit unnerving driving the wrong way up a freeway, but it was quick. In a few minutes they were at the Riverside freeway and up into and across the Anaheim and Chino Hills. They were safe. Homeless. But, safe.

———

February - Mexico City, Mexico

The airport was crowded. People were desperate to get to their destination before the disaster closed the airports on the seacoast. In addition, international flights to the many Mexican coastal resorts had all been diverted to Mexico City, adding to the crush.

The flight from New York had lifted off from Kennedy International that morning just as the first word from Antarctica was spreading around the world. The pilot had first announced to the passengers the full scope of what was happening mid-flight, when he also turned news broadcasts on the cabin speakers. Dee Dee Lindquist was stunned by the news. She had worked closely on the global warming issues and the Senator's legislation, so she knew more than the average person, but this news of the forthcoming global inundation was just as shocking to her as anyone else.

The trip through Mexican customs was the quickest such trip she could ever remember. A perfunctory look at her passport was all it took to get her through the inspection.

She was walking to the baggage claim area when she saw him. The white linen suit and sunglasses were out of character for him, but she understood the reasons for his Latin American dress. He was standing near the luggage claim ramps looking for her. It had been five months. She would surprise him.

Dee Dee carefully circled around out of his line of sight. He was watching the luggage ramps carefully and did not see her.

She walked up behind him, almost touching and whispered, "Hello stranger."

The man spun on his heels and faced her. Staring into Dee Dee's eyes for only a moment, he then gathered her into his arms and embraced her. Professor Joshua Corbell was more than Dee Dee's academic mentor

After a long moment the two broke from the embrace. Corbell spoke, " God, Dee Dee, it's good to see you."

"I got your message yesterday. I thought you were never going to contact me. I hadn't heard from you or Paul."

"No one has heard from Paul. Everyone else is accounted for, but he simply dropped out of sight. His explosions went off as planned, though, maybe he had an accident or has some reason to lie low."

Dee Dee shrugged. "Let's get my stuff. We can talk on the way to your place."

"Right, I suppose you heard the news?"

"Yes, on the plane. Can you believe it."

"Of course I can believe it. It is exactly what I, we, have been fighting against. It justifies everything we have ever done."

"I don't know about that. You know how I feel about some of it. What do we do now?" She paused to pull two bags from the ramp.

"I've got some ideas. The work isn't over yet."

The two walked out of the turnstyle at the entrance to the luggage area, only to find themselves surrounded by several men in the uniform of the Mexican Federal Judicial Police and two men in suits.

One of the men in suits opened a wallet and showed a badge with a five pointed star. "U.S. Secret Service... Joshua Corbell, Dierdre Lindquist, you are under arrest for the crimes of assassination of a member of Congress, flight to avoid prosecution and several counts of conspiracy. You have the right to remain silent, to consult with an attorney and if you cannot afford an attorney one can be provided for you, once we get back to the U.S., and anything you may say can be used against you in a court of law. Understand?"

One of the uniformed men pulled a card out of his pocket and read something in Spanish, his tone identical to the legal rights advisement of the Secret Service agent. As he did this, two other men pulled handcuffs from their belts and handcuffed Dee Dee Lindquist and Professor Joshua Corbell.

Agent Gabriel Marquez spoke again, "These gentlemen will be taking you to their office were a Mexican federal magistrate is waiting to sign your extradition papers.

———

February - St. Petersburg, Russia

The early morning sun brought little warmth. Commander Istvan Litasku stood with two other men, identically dressed in brown fur-trimmed parkas. They stood on the pier as the ice-breaking tug crushed its way up the lower Neva toward them. One of the men, his newly promoted second in command, Kolya Ivanovitch Tsverchenko, stood silently. The other man with Litasku spoke.

"I am sorry, Commander Litasku, I did not mean disrespect. I simply wondered the reason why we must have the patrol craft prepared and put into the water, it is highly unusual. Especially since only one can even navigate in the ice without the tug."

"Captain Shipilov, I have my orders, you have yours, That is how the system works."

"Yes, Commander." The surly detached attitude was unlike his commander, and Shipilov had been unsuccessful in gleaning the reason with his inquiry.

The large blue tugboat had now reached the pier and Shipilov left the other two to give instructions as to where the ice should be broken. Behind them the diesel engine on the large crane roared to life.

Two of the three patrol boats had been put on skids on the pier when the weather had turned so bitter in November. Even the one patrol boat with the reinforced prow had been pulled out when the thickness of the ice on the Neva and the waters of the Gulf of Finland exceeded its icebreaking abilities. Now, even though the ice was still thick in the River, little spring warming had been seen, the militsia commander had ordered the boats fully provisioned and back in the water. Tsverchenko had seen all the messages from Moscow that Litasku had, so he alone among the St. Petersburg police staff, knew what was driving Litasku.

Litasku turned to Tsverchenko, "Did your wife have anything to say about it? The preparations?"

"She did at first, but her father was military, so she can be a good little trooper when she needs to be."

"Then you are lucky. My wife, the intellectual, gave me a lot of grief, especially about not telling the hospital or anyone else until they, Moscow, give the word."

Shipilov returned to Litasku's side. The crane was now lifting the first boat into the water, even that behemoth of a crane was straining under the weight of the fifteen meter steel craft. Shipilov spoke, "The fuel trucks will be here shortly, I had a bit of trouble finding that much Number 2 Diesel with cold weather additive. Militsia vehicles do not use diesel so I got it from the city sanitation department. It is taking quite a bit. The KGB really designed these boats for the long haul, the whole keel and bilge areas are lined with fuel tanks. It will take thousands of gallons to fill all three full as you requested. But it will be done."

"Good, good."

Shipilov turned to Litasku. "Do you have any further orders? For after I have finished with the patrol craft?"

Istvan Litasku thought and then turned to Shipilov, his assistant for harbor enforcement. "Yes, there is. I want you to do a survey of what ships are in the harbor, both the inoperative ships and the operational ones who wintered over here or came in with the icebreakers. Figure out how many passengers each can carry under emergency conditions, and then work out a plan for boarding every vessel and seizing it in the name of the Russian Republic."

Shipilov laughed, but stopped when he saw that the St. Petersburg police chief was not smiling. Litasku continued, "And I want the patrol craft and your plan ready by this afternoon. I want the plan on my desk and I want you to tell Deputy Commander Tsverchenko how many men from the Naval Infantry Barracks you will need to assist in the seizure. And also, tell no one about the plan, until I tell you otherwise. Understand?"

Litasku was in his office that afternoon when the chime on the teletype out in his secretary's office went off. By the time he got there Tsverchenko was ripping off the yellow paper. Tsverchenko's face was white when he turned to Litasku.

Kolya Tsverchenko read, "From the Interior Ministry of the Russian Republic, to all stations. Execute `Yuzhnij Veter. ` Execute `Southern Wind.' Details to follow."

The muscles in Litasku's jaw tightened. He had gone over this a hundred times in his mind, now he must do it. "Kolya, notify the Mayor and the council leadership, and get Borodinsk, Petyev and Shipilov going. I will call General Tuibischev and the Admiral, I spoke with them earlier and they were waiting for the message also.

"Elena?" He turned to his secretary. "Elena, call down to my driver and tell him to carry out the instructions I gave him this morning. Make sure he understands. And Elena, I want you to know that one of his instructions is to pick your son up from school along with my family and Major Tsverchenko's wife and take them to a safe place. I have thought of you and your son, so that you can concentrate on helping me. Now, get the Director of Railroads on the telephone." Litasku turned and left Elena and Kolya staring at each other.

————

February - San Diego, California

"As we told you earlier, CBS Sports has been preparing for the PGA golf tournament in San Diego and part of that coverage included the use of the Metropolitan Life blimp for aerial shots. We are preparing to give you live coverage from the Met Life blimp over San Diego as soon as they give us the signal. But for now we would like to go live to Woods' Hole Oceanographic Institute where Dr. James Murdock is going to try and give us an idea of what is happening in the Pacific. Dr. Murdock."

The picture switched to a older balding man standing in front of a map of the world. The man paused for a minute while he listened to his earphone and then spoke, "Thank you. I suppose the first thing we need to make clear is that this is not your typical tsunami. A tsunami, erroneously called a tidal wave, is usually the result of an earthquake. A good example of this is the tsunami that struck northern California after the 1964 Alaskan earthquake or the 2011 Japanese earthquake and the great Indian Ocean earthquake and tsunami of 2004." He pointed to Alaska and then California, followed by the other locations. "Those tsunamis were caused by the movement of the ocean floor during the quake and they traveled thousands of miles at about 400 miles per hour and destroyed coastal property and killed residents in several places thousands of miles from the quake. Tsunamis are quite common in Japan with its active earthquake zone, hence their name.

"As far as we have been able to ascertain, this tsunami is the result of a major movement of the Ross Ice Shelf in Antarctica. We do not have any reliable information yet as to the extent of the movement, but it was enough to set in motion this tsunami which had a height of forty feet in New Zealand and Australia and somewhat less in Hawaii."

The announcer cut in with a question, "Can we expect the wave, when it hits the West Coast to be smaller still, as it appears to have lessened as it travelled across the Pacific?"

Murdock listened into the earphone and continued, "Not necessarily, although we'll know for sure in a few minutes. A tsunami's effects vary drastically by the type of terrain they pass over, that is, the ocean bottom. New Zealand and Hawaii have next to no continental shelf, whereas Southern California has a broad shelf which could allow the wave to crest out and be much higher and carry more water inland. Also, we only have word from a few places in the areas already hit, the size of the tsunami phenomenon could vary considerably by area.

"One further thing, although this tsunami is our current concern, everyone should realize that the real danger of this is the probability of sea level increase, permanent sea-level increase such as the President spoke of last March. If this major portion of the Ross Ice Shelf has broken off, we could be in for a sea-level increase of many feet. That is the reason for the evacuation warning for the entire U.S. shoreline."

"Thank you, Dr. Murdock, it looks like we are ready to go the blimp in San Diego."

The picture switched to an aerial shot showing coastline and a city in the background.

"This is Matt Cavanaugh, CBS Sports, aboard the Met Life blimp over San Diego. CBS wants to thank Captain Jonathan Teal and the crew of the Met Life blimp for this opportunity to bring you this eagle eye view of the situation in San Diego as the tsunami approaches.

"We just took off from Lindbergh Field in San Diego and we are passing over Coronado and Naval Air Station North Island, which you can see right now. In the channel to the right you can see one of the big aircraft carriers stationed in San Diego making for sea to avoid the worst of the tsunami near land. I understand that the tsunami effects are much worse along the shore than they are at sea."

"If we can move the camera down the coast," he paused and the view moved farther away, "now we can see the area of

Tiajuana to the south of San Diego. Since the tsunami is coming from the south we should see the wave first in this area."

"Captain Teal tells us that the radio reports the tsunami was just passed Ensenada which is what, twenty miles south, so we should see it in a moment."

Cavanaugh stopped talking while he and his audience waited for the wave to appear.

In the hazy distance of the television picture a white line appeared. As it grew in size it could be seen as a thin wedge, its narrow base touching the coast and narrowing out as it went out to sea. The camera refocused to full telephoto and the picture showed the hilly coast of Baja California south of Tiajuana. The wave was a white cloud boiling up the coastline.

"Now we can see some more detail of the tsunami. You can see to the right where the ridge of water is moving through the ocean and as it gets closer to land it curls into a classic wave top. I am not to sure of how high those bluffs are on the Mexican coast but the wave is nearly cresting over them. As we were leaving Lindbergh Field we got word that the civil defense authorities in San Diego were reasonably satisfied that they had evacuated all of the necessary areas of San Diego and its suburbs, I hope the Mexican authorities have done likewise, because this tsunami is wreaking havoc to the Tiajuana coastline."

"The large round building you see on the right of the picture is the bull ring near the beach area and it is also about the farthest building north in Mexico. When the wave passes that it will be in the United States."

Again Cavanaugh paused and they watched the wave roll northward.

"As you can see the ridge appears to be cresting into a wave farther and farther from shore as it progresses northward. That may be the coastal effect Dr. Murdock spoke of, or it may just be the perspective as we view this from a distance, moving in.

"There, wow, did you see that. That was the wave passing completely over the Tiajuana bull ring. It is now in the U.S. You can see as it leaves the mountainous shoreline on the Mexican side of the border it is spreading out farther inland. That first city you see is Imperial Beach. The wave is spreading to cover the entire town.

"The thin causeway you see now with the wave moving up it is the Silver Strand which leads to the city of Coronado. As you can see the wave is passing over that causeway as though it weren't even there. You can see part of the San Diego Naval Base coming into your picture now. Obviously they didn't get all of their ships out to sea.

"Oh My God, you can really get an idea of the size of that wave now. Look at it cover those huge warships. There look to the right. You can see one of the ships settling onto what I believe is Interstate 5. Ted, get the camera over on Coronado and downtown."

The camera panned to the left, now following the wave. "You can now see the wave passing completely over Coronado Island and the Naval Air Station. It is just rolling over everything. You can see downtown San Diego in the background.

"There, that building in the foreground is the San Diego Convention Center. Wow, the wave was up to the fifth or sixth story of the hotel. This is just the most amazing thing I have ever witnessed. I hope you folks watching can get some idea... My God, that skyscraper is going. It just hit the skyscraper and it fell right back over the other buildings. Look at the destruction. The whole of downtown San Diego, including the airfield we just took off from is under water.

"Jeez, Ted get over to the left, the aircraft carrier didn't make it." The camera panned to show the aircraft carrier, the USS Ronald Reagan, that had been steaming out of the channel was now pushed against a promontory of rocks at the harbor entrance.

"That huge carrier smashed like a toy on the rocks of Point Loma. Can we get the camera back on downtown?"

"As you can see, the tsunami has spent itself against the hills of Old Town San Diego and the water is flowing back down into the downtown area.

"We are going to continue later with our coverage here from San Diego, but now we understand that one of our affiliates has a helicopter with live footage from farther north up the coast as the tsunami continues its destructive march up the California coastline."

———

February - Del Mar, California

John Colton had, at first, intended to ride out the tsunami at Scripps Institute. But after the word of how high the tsunami had rolled in Hawaii, all the way up the slopes of the south shore of Oahu, he and the rest of the staff had decided to head away from the coast, as had most of the other people in La Jolla, even those on the bluffs. Besides, with three intervening low-lying areas on the freeway north to Solana Beach, which might fill when the water hit, it would have been impossible to get home to his wife and daughter afterwards.

Colton had filled the back of his old Land Rover with as much vital records as he could and had headed up the hill out of La Jolla. The evacuation had been many hours old and the streets were emptying when he left Scripps. There was some traffic near the university on top of the hill above La Jolla. Apparently they had decided the area was safe, as he had seen crowds of people, probably evacuees from lower areas, on campus.

The interstate had been busy heading north, but virtually empty heading south into San Diego. Although crowded, the freeway had been moving fairly well through north San Diego and Del Mar Highlands. Then things had started to slow.

By the time he came abreast of Del Mar Race Track, where he now was, the five northbound lanes had slowed to a crawl. Ahead he could see five lines of bumper to bumper traffic, all of the way across the valley floor and up into the bluffs of Solana Beach, and home.

He had the San Diego all-news station on and heard the reports of the passage of the tsunami past Ensenada. He was in the second lane and had his turn signal on trying to get into the right lane, but no one would let him change lanes.

His Land Rover had a sunroof and as traffic ground to a total halt he opened the roof and stood up to get a clear view ahead.

He could see the exit on the far side of the valley which he could take to the East, the back way, home. But, as he looked, he could see the smoke and flashing red lights at the exit ramp. It appeared to be blocked by an accident.

Del Mar Race Track had been built by the investments of movies stars in the '20's and '30's. The coastal slough and tidelands had been drained and the race track and surrounding stables and parking areas had been built. It was only a matter of minutes from his home and Colton had regularly came here to jog down to the beach.

Being thus familiar with the terrain, Colton new that the entire tideland valley was right at sea-level. In fact, big storms often swamped the beach houses on the opposite side of the race track. This was not the place he wanted to get caught in the tsunami.

The radio now announced the first word of the wave hitting Tiajuana and San Diego's southern suburbs. Traffic continued to stand still. Horns blared uselessly. People started to get out of their vehicles. Behind him in his rearview mirror he could see hundreds of drivers, leaving their cars and start running back up the freeway towards the Del Mar Highlands.

The radio announced the destruction of Coronado. Many cars in the right lane started to drive off of the freeway, down the embankment and off across the horse pasture that bordered the freeway. When the two cars next to him took this route, Colton, too, made his move.

Cramping his wheel and gunning his engine he jerked into the vacant space beside him before the cars behind could pull up. He steered himself off the freeway and down the bank.

The two cars ahead of him had already knocked the freeway's chain link fence and white wooden fence of the horse farm down. He followed their path through, but had to turn quickly to avoid hitting them. Both were stuck in deep mud in the pasture. The torrential rains that had drenched southern California for months had turned the pasture into a quagmire.

He was still rolling when he put his clutch in neutral and slipped the vehicle's four-wheel drive into Low 4. The gearbox clunked into place and he continued to move.

The radio announcer had been cut off in mid-sentence by silence. The downtown San Diego station went off the air. Colton now relied on instinct to guess how much time he had. He was barely half way across the pasture, churning mud, when he saw the wave in the rearview mirror. It was too late.

He thought of making a run for it, but it would have done no good. It was upon him.

Colton stopped and stood up in the sunroof just in time to see the mighty wave crash into the western end of Solana Beach, To the left the grandstand at the race track, in the shadow of the Highlands and thus out of direct line of the tsunami, was nevertheless folding down under the weight of the water that now flooded in from the beach area.

He could hear the roar of the wave as it crashed into the town on the hills to his right. It was like the regular surf, but a hundred times louder. The force of the wave hit the cars across the valley near the exit first. Colton saw them flushed aside like toys.

As the main force of the wave dissipated back across the valley, the rolling waters behind it flowed into the tidelands. Without much time to think of whether to stay in the car or not, John Colton climbed up through the sunroof and stood on the roof rack. In seconds there was a rush of water over the cars and running people nearest him on the freeway. With a roar the wave engulfed the oceanographer in the cold waters of the Pacific Ocean.

———

February – Florida and Southern Georgia

Guillermo "Billy" Barraca was rapidly leaving the known world. Or at least the world known to him since his youth. He had absent-mindedly noted as he passed through Orlando, that the farthest north he had ever gone was that one trip to Disney World as a teenager. Now as he drove the Cadillac over the Georgia state line and onward through someplace called Valdosta he was truly out of his home turf.

He looked over at Lucy. Lucy? Yeh, that was her name, Lucy. She was curled up in the passenger seat asleep. His sport jacket that he had put on her when she had complained of being cold was now piled up around her arms and shoulders. The black velvet dress was all balled up around her waist showing garter belt and stockings. In the dim light of the dashboard and oncoming traffic she did not look nearly as old as the twenty-one she claimed to be. It was amazing the maturity that came with sexy clothes and a well applied coat of makeup. Whatever it was, this Lucy was just what he liked, young, pretty and built.

It was pure circumstance that they were together on this journey. He had only been with her for one night before last night and now they were fellow refugees. He had picked her up at the same haunt she had been in when they first met, a nightclub called the Delphinium. She had earned her money. When the heat of their passion and the powdery chill in their nostrils had finally ceased just before dawn, they had collapsed in each other's arms and slept until mid-afternoon.

He had awoke to find that the FM station they had been listening to throughout the night was still on, but playing an endless loop public service announcement of the warning to evacuate Miami. He listened incredulously to the announcement and then ran over to the balcony. He could hear traffic on the causeway to the north, but Biscayne Boulevard below him was strangely vacant. No pedestrians and just two vehicles, both heading north.

He went inside and awoke the cute little hooker, Lucy. In her post-"coke sleep" stupor she could only listen to the tape forever repeating warnings and evacuation routes and mutter, "Would you listen to that?" again and again.

Guillermo noticed that his phone on the bedside table was off the hook, he could not remember whether it was intentional or not. He cradled the phone and put it on speaker as he packed his essentials. His phone was the type that stored common phone numbers, which he cycled through, trying to find someone in his circle still in Miami. Neither friends, family nor "business associates" answered.

Lucy finally revived enough to start worrying about herself. After a thought of dumping her, Guillermo assured her he would take care of her.

Seeing him packing his bags, she asked, "Can we go get my stuff?"

"Where do you live?"

"Coral Gables. It's right on Dixie Highway."

"Shit, Miami is fucking flooding any time now and you want me to drive way down to Coral Gables. You want to go to Coral Gables, you go. My bus is going for the high ground."

Lucy dressed and then sat on the bed blinking her eyes at him, trying to comprehend her situation. Guillermo continued packing.

"But, Billy..."

"Look. Go into the extra room, there is a bunch of shit my ex-girlfriend left, I never got around to getting rid of it. You can probably find some stuff you can use. And it is probably better shit than you own anyway."

Lucy went into the other room. As she left Guillermo snapped his fingers and muttered, "Shit, I almost forgot" as he ran to the kitchen.

In the pantry closet, Guillermo found the bulky attache case, the kind the attorneys for the cartel boss always carried to court. It was hidden under the foodstuffs, waiting for his man Raul to take it for their next purchase. He took it to the dining table and snapped it open. The dozens of stacks of used twenties and hundreds were neatly arranged. Guillermo fingered them and wondered aloud, "Where the fuck is Raul, anyway?" None of his men or the boss had contacted him. Perhaps it was every man for himself today. Well, he could easily deal with the problem of being by himself when it meant having to keep this valise and its contents.

He heard Lucy set two suitcases down behind him. He turned to see her staring wide-eyed at the contents of the valise. He snapped it shut and said, "Just some traveling money."

"The suitcases in the bedroom were already packed." Lucy said questioningly.

"Yeh, well, let's say Angelica had to leave rather quickly. Circumstances and the D.E.A. required her to take an earlier flight back to Columbia than she had planned. Her stuff will probably fit you, but it might be a little tight on top for you." He finished the last word off with a stiff-fingered jab at a nipple, which she fended off as only a girl who works the bars can.

"Let's get the fuck outta here!" Guillermo said as he hefted the case. Lucy had grabbed her new suitcases and followed him to the door. He grabbed his other bag and they were off.

They loaded his Cadillac with their bags in silence. The parking garage was already empty. He wondered when the first call had gone out on the evacuation.

Now he drove on through the Georgia night. The traffic had eased as they went further north. Traffic jams had been worst around Orlando and again as the connections to Tampa and Jacksonville emptied into their route, but now in the middle of the night the traffic was not much worse than any freeway he was familiar with in Miami.

Guillermo took stock of his situation as he drove. He knew absolutely no one outside of Miami, at least not in the U.S. He had no idea of where he was going. The only world he knew was scheduled for destruction any time now. He had a good car, a decent looking woman and a hell of a lot of cash. Things could be worse.

———

February - Garo Hills, East Bengal

The first radio reports came in just before midnight. There had been confusion about what was happening. The first telephone call from Mymensingh had spoke of a flood. The radio from Calcutta spoke of a tidal wave, before it went off the air. By the time his men had roused Darpheel and he had gone to the two-way radio in the Land Rover he could not raise Dhaka.

He had been spending a long weekend at home in the Garo Hills with his family, Latifa, and her children. He took every opportunity he could to try and escape the squalor, refugees and constant responsibility which faced him in Mymensingh and the other cities in his region. The state of affairs was such that he could afford to take some time away from his duties and he did so whenever possible. Majlis was getting things well in hand in Dhaka and that relieved pressure on Darpheel and Majlis' other assistant military governors in the outlying areas.

And now this. Darpheel was unable to comprehend exactly what was going on, but there was apparently flooding in Dhaka, Calcutta and much of the coast. He had, of course, read the articles about the possibility of coastal flooding caused by the global warming, but he refused to consider this possibility as he loaded the Land Rover and headed out of the mountains. He would be in Mymensingh by early morning and find out what was going on, using the military communications set at the headquarters there.

As it winds down to the lowlands, the road out of the Garo Hills circles around above the railroad bridge and coalyard where Akim had first spoken to Latifa . It was on this last outcropping of high ground on the road above Mymensingh that Akim Darpheel got his first look at the new world.

He had seen the same majestic view from this outcropping every time he had gone into the city since he was a young boy. Now it was changed beyond recognition.

Normally, you could see from Mymensingh on the right where it was pushed up against the low hills, all the way out across the lowlands that had been the heart of Bangladesh. This land, which had been formed by the silt of the Ganges and Brahmaputra Rivers, was, but for the hills of his own home, virtually flat, just as the river delta had left it eons before.

Now, in the early dawn light, he could see some of the lights of Mymensingh to the north west, but to the south and southeast where he should see the fields, farms and towns of the countryside between him and Dhaka he could see only the glistening expanse of water, punctuated in places by rows and clumps of trees, and an occasional rooftop.

The limit of the water was unclear, but it seemed that Mymensingh was on dry land. Darpheel and the men with him, both Indian and Bengali, got out of the vehicles and went to the edge of the road to look out over the valley and delta area below. A shudder passed through Darpheel as he watched silently. He could not help but think of the journey by boat out of Narayanganj so many months before. Now, it seemed, the entire country had taken on the visage of that watery hell.

As the sunlight grew stronger, Darpheel could see the first group of refugeesbelow him in the distance. They were crossing the railroad bridge at the foot of the Garo Hills as he once had, trying to reach the hills from the flooded wetlands. Something would have to be done, for them and the tens of millions more who might follow, if they still lived.

Thinking first of Latifa, he knew that the men he had stationed in his home village would take care of her and the children for the time being. He would send word to her when he could. For now, he would be needed in Mymensingh. He had to check on the extent of the flooding. Had Majlis survived in Dhaka? What had become of Dhaka?

Brevet Colonel Akim Darpheel and his men got back into the vehicles. He ordered his driver to take the old high road into Mymensingh, the main highway went too close to the river. There was work to be done.

———

February - San Bernardino, California

"How much?"

"A party of four will be two hundred fifty seven ten with tax."

"Since when? This motel is a national chain with published rates. You can't just change rates because of the disaster."

"We can and we did. Special events are in our franchise contract. And the owner considers it a special event when half of Los Angeles is moving inland looking for someplace to stay." The young kid behind the counter had obviously been instructed on what to say, or he had already said it a dozen times. "Now. I've got four rooms left and four people behind you in line. Do you want to take it or not?"

Ian hesitated a moment and then pulled his wallet out. He hated to be taken advantage of. And that was what was happening here. Highway robbery, quite literally. But, he had checked at seven other motels between Riverside and San Bernardino and this was the first with openings. He signed his name and harrumpfed at the motel clerk's automatic "Have a nice stay."

Ian Petersen took the key and walked to his car. He dangled the key in the air for Cyndi, in the other car, to see and she smiled a smile of relief which he did not return. He got in his car and drove around to the side of the motel where their room was. Cyndi followed in the Volvo.

Ian and Crystal were already in the room when Cyndi and Amy got there, having had to park a bit farther away.

"Well, this isn't too bad. And it overlooks the pool," was Cyndi's reaction to the room.

"For two hundred fifty seven bucks for a fleabag motel it had better be fucking marvelous," Ian's Canadian accent always came out strongly when he swore.

"What?"

"Yeh, the motel owner scalped me clean and I watched with my eye's open. Go figure."

"But isn't that against the law?"

"I don't know, it may be. But knowing that it is wouldn't get us a place to stay for the night. We'll see how thing go and maybe find someplace else for tomorrow night."

Amy spoke now, "Tomorrow night? When are we going home?"

Ian walked over and put his hand on her head, "I don't know, Hon. Let's turn on the TV and see if we can find out how things are going."

The room was starting to get messy as only a motel room with a family staying in it can. Wrappers from the delicatessen foods and empty soda cans sat around. One of the beds was covered with clothes where Crystal had tried to sort through the things Cyndi had packed before declaring that her mother had not packed 'any' of the right clothes.

Cyndi and Ian sat on the end of the other bed with Amy napping behind them. Cyndi's eyes were puffy from tears and Ian had his arm around her. The television was still on as it had been for several hours.

"God, did you see that!" Crystal finally said. None of them had said anything since television pictures from the Los Angeles station's traffic helicopter had started. "Look, they've got surf rolling on the 405 freeway."

"I just can't believe it. The only thing I could recognize in Newport Beach was the big overpass sticking out of the water and the highrises weren't even there." Ian punctuated his words with the television remote trying to see if any other station had something different. He stopped at the cable news channel which was showing the aerial shot of the tsunami coming up the coast again, they had seen it several times already, but watched it again, as though new, so enthralled were they at seeing their home and community disappear beneath the rolling wave.

When the news channel switched to a prediction chart for when the high sea levels would reach the East Coast, Crystal turned to her mother and step-father and asked the question that had gone unspoken so far. "What are we going to do?"

"Maybe we can rebuild." Cyndi suggested.

"Are you kidding. You saw that last shot from the chopper, the damn beach is going to be two miles in from where our building was. And rebuild with what? Most of the coast and entire cities like Long Beach are gone. They'll be lucky if there's one insurance company left in business."

As Ian stopped talking he saw that his harsh assessment had started Cyndi's tears rolling again and Crystal was now close to that also. He stopped and thought for a moment. Then he got up and walked over to the suitcases, talking as he went. "However, there is some good news. Mother Nature appears to have gotten me out from under the mortgage on the apartment house I couldn't sell. If I can't get money out of the insurance company the mortgage company can't get it out of me."

Ian pulled his gym bag from the pile of luggage and brought it to the bed. He carefully held the 9mm pistol with one hand to keep it from tumbling out as he dumped the contents of the bag on the bed. "Why don't you ladies count our working capital while I take a shower." Ian knew exactly how much the three days receipts from the restaurant and the monthly rents deposit he had dumped out was, but he figured that counting the thousands in cash, checks and credit card slips would take his girls' mind off the troubles.

"I propose that we use this to get settled on the land in Nevada."

"That old trailer?" Crystal seemed repulsed.

"Yeh, the trailer. And a hundred fifty acres of fairly nice development land. We can stay there until things calm down a bit. Then we can use this and the money in the bank in Nevada to start out new somewhere. Maybe somewhere a long way from the ocean."

As he looked down at Cyndi start to count the money he wondered momentarily whether the checks and credit card slips were still good. He had heard the President announce a state of emergency on the television only an hour before, but the President had not spoken of such things. But Ian wondered how you could cash a check or use a credit card from a bank in Newport Beach or San Diego which was not there any longer. Did he have any right to cash rent checks for an apartment that was under water now? And how could the credit system still work when half the state of California was demolished or the residents jobless or homeless.

With this thought about the credit card system Ian had a curious sense of satisfaction. He had paid the young scalper at the motel desk with a credit card from the bank in Newport Beach; the bank that had not been visible in the pictures from the television traffic helicopter just now.

———

February - St. Petersburg, Russia

It was the early morning hours of the next day after the "Southern Wind" notification had come in when Litasku's black Zil pulled up on the pier again. He arrived just in time to see one of the patrol boats and the icebreaking tug load a group of people on board and pull away to some destination in the night. The passengers appeared to be working class families, bedrolls and canvas suitcases in hand, probably to be billeted on a freighter tied up someplace in the dark harbor. The other two of his patrol craft were tied to the pier. A group of people who he recognized as some of his junior staff members and their families huddled on the pier, apparently waiting for the last patrol boat to return for them.

Shipilov was standing on the pier, apparently having just dispatched the load of evacuees on the other boats. There could be no other reason to be out on the pier at this hour, in this cold. As he walked up, Shipilov saluted Litasku. "Commander, the waterfront is evacuated and your staff is in the first craft, the reinforced one. I will be in the second one here, the Deputy Mayor is aboard also, as are several other important, err, ahh, persons."

"Very well, get inside yourself. And prepare to get underway as soon as the other boats come back in." Litasku saluted back and headed for the first boat. His driver carried two suitcases which he handed to a fellow police officer on the deck of the boat below them.

Litasku entered the door of the cabin and was relishing the warmth and getting his eyes adjusted to the light when he heard a squeal of delight and felt a hug around his hips.

"Papa!" Daine had found him.

Sofia also gave him a quick hug. There was not much room to move in the cabin which had probably twenty people in it, including women and children. By the time Istvan had taken off his greatcoat and given it to Sofia, he noticed Kolya Tsverchenko at the entrance to the next cabin, the radio room and command center. He made sure his wife and daughter were settled and followed Kolya into the command center.

The first thing Litasku noticed in the next room was his father, Nikolasha Litasku, still in coat and scarf bending over the navigation table marking on a chart with protractor and straightedge. Another six to eight of his officers crowded into the room. Walking up to the table the militsia commander noticed that the chart was of the Baltic Sea and Gulf of Finland. His father was drawing expanding crescents on the map centered off the southern tip of Sweden.

"I see you are making yourself useful, Father."

"Your police officers did not know which side of a nautical chart was up, so I thought I would help out."

"And what have you found?"

"If the word from Scszecin is correct you have about an hour, maybe an hour and a half, before the water comes up. And, it appears there are several distinct surges, nothing like the tsunami that hit the Pacific, but a lot of water coming in with the tides."

Kolya was now at Litasku's arm and spoke, "We had word on the shortwave set from Copenhagen when the surge passed through the Kattegat, but they went off the air shortly thereafter. The last word we have had was from a Polish freighter leaving harbor and that is what your father plotted out."

Istvan Litasku turned to his chief deputy, "And what word of the evacuation?"

Kolya pointed to the map taped to the wall behind them. "The bulk of those who went by train appear to be safe on the rail lines to the south and east, of course, that assumes that the inland cities are safe. We don't have decent charts for that kind of thing, but I understand that the marshes around Novgorod really aren't that high above sea level. All we can do is pray the water isn't too high."

"And the others?"

"Well, there was a huge traffic jam heading north to the elevated forest areas, but there isn't any services at all in the forest, just two little towns. Mostly dachas and cabins. They are going to be in dire straights there in this cold, but it was better than drowning. I guess. And also, one of our units heading north into the Karelian forest was met by a convoy coming out of Vyborg, apparently they are heading to the high ground in the forest, too."

"That would make sense. Vyborg, Petrozavodsk and the whole countryside way out to the White Sea on the canal are fairly flat and barely any elevation. Lake Ladoga is only a few feet above sea level." Nikolasha interrupted Kolya and thumped the chart near Ladoga for emphasis.

Tsverchenko continued, "And in the southwest along the rail line, General Tuibeschev had some trouble with the Estonian border guards. I don't know why they had to go all the way to the Estonian border, but... I guess Tartu is one of the few cities within a reasonable distance with high ground. Tuibeschev finally convinced the border guards that for humanitarian reasons the border should be opened to our refugees trying to take the train. But I understand most of them have located in the camps Tuibeschev set up just south of the rail line and highway between the city and the Estonian border. Here and here." Tsverchenko pointed the chart.

Istvan smiled, "I can imagine Major General Tuibeschev's discussion of humanitarian principles with an Estonian border guard. Humanitarianism by gunpoint." Thinking about what he had said, Istvan quickly turned to his father and with a solemn shake of his head and stern look, convinced the old Estonian to hold his tongue.

In the darkness outside the boat's cabin, they could hear the rumble of the diesel engines of the third patrol craft and the icebreaking tug as they returned to pick up the last of the staff's families on the pier.

Kolya continued his report, "Last reports from the remaining patrols in the city are that only a few cars and buses are still leaving the city and, except for a few muttonheads who won't leave and the ones we put up in the highrises, the city is empty."

Istvan studied the map on the wall. He touched the one clump of darker green to the north of St. Petersburg which indicated the high ground nearly a third of the city had run to. The egg-shaped blotch of ground was centered on the isthmus between the sea and Lake Ladoga, halfway from St. Petersburg to the border with Finland. "So, we have evacuated a city of millions and dumped them in the middle of the frozen forest at night without provisions. Do we count that as a success?" He raised his eyebrows at Kolya to end the question.

Kolya shrugged, "Moscow acknowledged our last request by radio while you were on your way down here. They understand the situation and may be able to parachute some help in to both the northern and southern areas. They acknowledged that we need tents and field kitchens and, of course, the blankets and field clothing. But they don't know how far the water will go inland and they cannot commit to aid before Thursday and possibly not then."

"Yes, well don't hold your breath," Istvan spat out the words. They all knew the horrors facing the citizens of St. Petersburg in the Karelian and Estonian forests. That they, themselves, were safe and warm on a sturdy ship went unspoken.

"And how about the ones on the boats?"

"Shipilov did his best, but we could only put a few thousand on the boats. There just isn't that much sea traffic in St. Petersburg in the dead of winter. The Navy did take care of all their families and plenty of others also."

"Well, at least they will be better off than the poor bastards in the forests."

Istvan had one more question for Kolya, "And Petyev?" He spoke under his breath and touched his lip with forefinger.

Kolya answered in a hushed tone, "He was as successful as could be hoped. He has several members of the curator staff from the Hermitage and most of the greatest art treasures on board the tug. They have a reasonable sized covered hold area, and he took command of the tug himself."

"Khorosho. Very good."

Istvan turned to the lieutenant in charge of the patrol boat's crew. "Signal the other militsia craft to follow us out of the harbor. Let's go about two or three miles out beyond Kronstadt and wait there to see how this goes. We don't want to be too close to shore when the surge comes in."

"Aye, aye, commander," said the young officer as he turned and climbed the ladder to the bridge.

It was only with the jaunty nautical response of his junior officer that Istvan Litasku recognized that he had, with that order, ceased to be the police chief of St. Petersburg and had become the commander of a small naval squadron, afloat on a very cold and rapidly rising ocean.

———

February - Solana Beach, California

The marble fireplace was meant for decorative purposes, not practical ones. But this evening, since every appliance in the house and the heater required electricity, the logs in the fireplace were serving a useful purpose.

Electricity had been out since early afternoon. Another of the now frequent thunderstorms had dropped the temperature. She had moved the baby's bassinet over within the fireplace's warmth and she had brought a futon and blanket over next to the bassinet to lay on. A bottle of formula and a jar of strained fruit warmed on the marble in front of the fire for the baby when she woke.

The telephones were out. She had listened to the portable radio for hours until the baby started to fall asleep. There was morbid, horrible news from everywhere, statewide and worldwide, which did nothing to excise her greatest fear.

She had heard of the imminent disaster when she had arrived at the home of a real estate client in late morning. When she got home from getting the baby at the nursery there had been two messages on the phone recorder. The first from her husband said he would be staying at work and that he would call again soon. The next call, around noon, had said that he would be coming home after all and for her and the baby to stay home where he said they would be safe.

That had been eight hours before. She had neither seen nor heard anything since. From anyone. Except the radio and its messages of destruction and death.

Now she sat alone in front of the fire. She tried to tell herself that no news was good news, but since she was cut off from any news, good or bad, of her husband this was little solace.

The baby cooed in her sleep and stirred in the bassinet, but did not wake. Far away in the night she heard a siren, the alternating tones of an ambulance or perhaps the paramedics. That did not help her any. It just set her worst imagination to work.

She got up and pushed the baby's bottle and food back from the fire a bit. She walked to the bathroom.

Half way to the bathroom she saw the pattern of light on the dining room ceiling that she knew was shining through the front window from a car coming up the hill. She waited in the hall while the light grew brighter. When the pattern of light changed to show it turning into her own driveway, she turned and raced to the front door, flinging it open.

Her heart stopped when she saw the markings of the Highway Patrol on the car. *Weren't they the ones who notified the families of accident victims?*

For a moment that wore one as an eternity she stood transfixed in the glare of the car's headlights. Then, the passenger side door opened.

A tall figure in a long black robe got out of the car. No, it was a blanket, not a robe. She gave an involuntary yelp of joy when she recognized the car's passenger.

Rebecca ran and threw her arms around a muddy and wet John Colton. They said nothing, just hugged, as the Highway Patrolman backed away and drove off.

———

February - Off the New England Coast

VIA USCGCOMMFAC - 1 PAGES XMITED
S E C R E T (XGDS-2)
FM: CMDR COHEN, USCGS BRISTOL BAY
TO: GAIA OFFICE, NCAR, BOULDER, CO
INFO:SCRIPPS INST,(ATTN: DR. COLTON) SAN
DIEGO, CA
GAIA DISTRIBUTION LIST ECHO
NSC, WHITE HOUSE, WASHINGTON, DC
SUBJ: SITREP OF SIGNIFICANT MARITIME
INFORMATION - FEBRUARY(PARTIAL)
1. THIS WILL BE THE LAST MARITIME SITREP
UNTIL FURTHER NOTICE. USCG/GAIA OFFICE
IN NEW LONDON EVACUATED DUE TO EXPECTED
COASTAL FLOODING. STAFF REASSIGNED FOR
OPERATIONAL NECESSITY. CDR COHEN HAS
EMBARKED ON THE USCG CUTTER BRISTOL BAY.
2. REPORTING STATIONS WHICH REMAIN
IN OPERATION IN THE ALREADY AFFECTED
AREAS INDICATE A SEA-LEVEL INCREASE OF
BETWEEN SEVENTEEN AND TWENTY-FIVE FEET
WORLDWIDE, DEPENDING ON LOCAL TOPOGRAPHY
AND LATITUDE. NORMAL LUNAR TIDES HAVE
APPARENTLY CHANGED IN PROPORTION TO SEA-
LEVEL INCREASE.
3. FRENCH NAVY BULLETIN REPORTS MASSIVE
"RED TIDE" IN MEDITERRANEAN SEA BETWEEN
MAJORCA AND SARDINIA. AREA OF OVER FIVE
HUNDRED SQUARE MILES. "RED TIDE" IS AN
INCREASE IN ANAEROBIC BACTERIA/ALGAE IN
SEAWATER WHICH CUTS OFF OXYGEN TO NORMAL
SEALIFE, RESEMBLES RED SEAWATER WITH HUGE

FISH KILL. PHENOMENON USUALLY THE RESULT
OF HIGH WATER TEMPERATURE.
4. GOOD LUCK FROM COHEN TO GAIA TEAM.
5. THIS REPORT IS CLASSIFIED IN
ACCORDANCE WITH EXECUTIVE ORDER 14708
AND INFORMATION IN THIS REPORT MAY NOT
BE RELEASED TO THIRD PARTIES WITHOUT
AUTHORITY OF NSC, WASH, DC.
XGDS-2
XXX

February - Washington, D.C.

Andy Knowles poked the power switch on his office television. He had seen enough. More than enough.

Like many of his countrymen and people around the world, the Senator had watched death and destruction for several hours. He had bemusedly watched the President announce the state of emergency. The situation was such that it was impossible to have an 'I told you so' attitude about it. Sitting in his office, Andy had gone over everything in his mind time and again.

It was not like there was anything they could have done. At least, not at this late date.

If they would have had the ten or fifteen years Andy had spoken of to the senators and congressmen that day in the Greenhouse Gas Act compromise meeting, perhaps they could have done something. But this? Less than a year from when Brady had brought the GAIA information to him. And only six months since the first warnings from the seismologists.

He stopped himself. This was stupid. In the shock of the magnitude of the catastrophe that was occurring, he was making excuses. Useless excuses. Worthless excuses.

Belatedly yearning for ten or fifteen years to do something worthwhile to stop the global warming was a denial of the truth. If anyone had wanted to listen, there had been plenty of warnings for two decades or more now. And even after the GAIA team had shoved the facts down their throats last year, those in power had been unable to make the hard choices and take action to right the century and a half of industrial pollution which had set this day of catastrophe in motion.

And Senator Andrew Knowles was one of those in power. Like his compatriots elsewhere in government, he had knuckled under to political necessity, had compromised on environmental goals, had idolized economic development, had compromised for jobs, had flirted with public opinion, and apparently in the end, had compromised the health of the planet. In going over the myriad of compromises that he, himself, had formulated in his years on Capitol Hill, Andy wondered if in those compromises lay the guilt for this worldwide devastation. Were those who were concerned with the environment and compromised themselves to achieve part of their goals or an interim measure more guilty than those who neither cared nor believed? Andy assumed that questions such as this fell in the 'history will be the judge' category.

The thought of 'guilt' and 'compromise' made him think of his own daughter. Like most of those in the New Jersey demonstration, she had been released without charges being pressed. The New Jersey prosecutor had aspirations of higher office and recognized the environmental movement as a big block of his party's constituency. But, Andy had taken the opportunity of Sandy coming home the next weekend after her arrest to have a long talk with her about the environmental cause she had risked arrest for.

When Andy had heard his daughter speak of rain forests and greenhouse gases, he had heard his own words. But he heard his words without the discolorations left by politics. When Sandy spoke of a 'sustainable economy' and 'species diversity' he knew she believed in these things which he had spoken of a hundred times in speeches, articles and political debate. But when she spoke these words her belief in them was untainted by the art of compromise and political reality. Sandy had spoken about the environment and the scientific principles in clear and simple terms. In her eyes what was beneficial to the global environment was good and what was not was bad.

Andy wondered how the lunchtime arm-twisting with old Senator Hudspeth last May would have gone if Andy had shown the youthful exuberance and unequivocal belief in what was right that Sandy had expressed to him when she spoke of clear-cutting forests. It was an interesting thought, but Andy knew the answer. At the time he had spoken to Hudspeth, the old senator had held the trump cards, money and political advantage.

Well, the old man from Oregon had gone on to whatever reward awaited him in the hereafter. And those still living on this planet were today getting their reward for letting men like Hudspeth stay in power for too long.

Andy needed to get out of the office. He felt the need to do something, but there was nothing to be done. After he had gotten the first word early in the morning, Andy had called to insure that Sandy was on her way home from New York. A late afternoon call from Molly had confirmed that she and Sandy were safe at home in Bethesda and a call to the city had confirmed that none of Bethesda was in any real danger of flooding from the tidal wash in the Potomac River.

The rest of the office was empty when he left it. It was late. Checking his watch Andy noticed that it was only thirty or forty minutes until the time when the high water was due in the Chesapeake Bay and Potomac River area. Andy knew where he needed to be.

Parking on the lower Mall was for once easy. Parks Police and the D.C. cops were blockading the area and it had only been his congressional license plates and the insistence of this well-known senator that had persuaded the cop to let him through.

He parked near the Bureau of Engraving and Printing building. The night was clear, not too cold and a full moon was out.

Andy thought of going down to the Lincoln Memorial near the river, but his recollection of the area surrounding the Lincoln Memorial led him to select the Washington Monument as his goal. It was right across the street from the Bureau of Printing building and the high rounded knoll upon which it sat would give him a good view of everything, from the Jefferson Memorial to the Capitol, the White House and, most importantly this night, the Tidal Basin and, beyond the Jefferson and Lincoln memorials, the Potomac River.

The police had done their job well. There was not a vehicle moving on the Mall from the Capitol to the State Department. Beyond the Lincoln Memorial he could see that the Memorial Bridge was closed, too. All of the low-lying area in the city had been closed off from the public.

In a city that was as constantly active as Washington, it was eerie to be walking in virtual silence. There were not even any planes taking off from National Airport. But then, there would not be any. Not tonight, perhaps not ever. National Airport was built on the banks of the Potomac, barely a few feet above the river. And tonight the Potomac would reclaim it banks.

Today's second White House briefing, that he had attended just after lunch, had included an estimate of when and how high the Potomac would rise. Being slightly higher than sea-level, but not much, the Potomac River was predicted to come up twelve to thirteen feet, perhaps more later when the rest of the ice shelves possibly followed the Ross Shelf. It was not as much as would hit the cities right on the ocean, but it would have a profound effect.

There was a light layer of snow on the ground that had partially thawed during the day and now had re-frozen in the brisk night air. It crunched underfoot as Andy made his way across the lawn and started up the slope. Andy reached the concrete protective barriers which surrounded the Washington Monument. He was heading for the southwest side of the knoll. Above him at the base of the monument near the ring of flagpoles he could see a group of people, a few of them in the floodlit glare that illuminated the monument and some whose shadowy shapes he could make out on the same southwest corner he was heading for. Obviously someone else had the same idea as the Senator.

He had only covered half the distance from the barricades to the top when one of the shadowy figures intercepted him. The man had trotted down the hill toward him and slowed a few yards ahead of him. He was a young man in a bulky dark parka.

The man held up one hand, signaling Knowles to stop and said, "The monument area is closed. You'll have to leave the area."

Andy stopped short. He could just make out the man's features in the moonlight. He started to argue the point and explain who he was, but stopped. The man had moved directly between him and the group at the top of the hill, putting the man's dark silhouette against the lighted backdrop of the monument. That silhouette had a small wire coiling from ear to collar.

Obviously, someone else at the White House briefing had been intrigued by the prediction about the Potomac rising. Someone with enough clout to have a Secret Service cortege seal off the Washington Monument grounds so that he could watch the event take place.

Andy spoke, "I'm Senator Andrew Knowles. I have White House clearance. My clearance badge is in my suit pocket if you need to see it."

A penlight snapped on in the agent's hand, illuminating Andy's face.

"Your ID won't be necessary, Senator." The agent then spoke something into his coat sleeve. Andy could not hear what was said.

There was a pause, while the agent touched his ear. Finally the young man stood aside and motioned with his arm for Andy to proceed up the hill.

Andy continued up the knoll to the monument, circling up to and around the ring of flagpoles to where the group stood. At last he could recognize the figures in the group.

The President's penchant for jogging to relieve stress and reduce his paunch was notorious. In fact, it was the target of quite a bit of humor and, reportedly, complaints from the Secret Service agents thus frustrated in their efforts to limit security risks to their important charge. The President had chosen to unwind after the national address on the state of emergency with this trip to the Mall with his close friend and advisor, for Andy could now see both the President and the Vice-President standing amid a ring of vigilant statues who faced outward. The ring parted and admitted Andy to its midst.

Noting Andy's arrival, the President turned and stretched out his hand. "Good evening, Senator. Great minds, Eh?"

"Good Evening, Mr. President."

The Vice-President also reached out to take Andy's hand. "Andy? Good to see you."

Andy nodded and asked, "Any word? On when it is coming?"

Both of them started to answer, but the Vice-President deferred to the President. "Agent Floyd has a cell phone line open to the Park Service at Mount Vernon. It has just started there so it should be any minute."

The three stood in silence. Around them, in an uninterrupted vista, was the city of Washington. The moon was shining full in a cloudless sky. To the south was the dome of the Jefferson Memorial and behind and to the right of it the waters of the river and the Tidal Basin. Beyond that the lights of National Airport and the Virginia shore of the Potomac. Down the hill from them, between them and the Ellipse with the White House beyond was a string of cars that Andy had not noticed on his trip up the opposite side of the hill. Two limousines and two sedans sat with their lights out. The President's emergency transportation and the car carrying the warrant officer with the communications gear and nuclear codes were never far behind the President's jogging group.

As the silent vigil continued a curious and bothersome thought struck Andy. Here he was, standing next to the man who was the most powerful man on Earth. And this man who commanded power and respect worldwide stood, like Andy, as a helpless spectator as the waters rolled in towards his capital city. They were watching like the billions of other human souls as Mother Nature, Ken Brady's Gaia, wreaked her revenge for the Industrial Revolution and the myriad of other insults mankind had exposed the planet to.

And there was the incongruity of the President and Vice-President standing here in casual attire while this calamity struck the country. But then, having the President standing in the Situation Room wringing his hands would serve no purpose. Nothing more could be done. Not now. And this was the President's style. He liked to participate and feel the events and situations he faced. The forlorn concern Andy saw on the President's face in the moonlight was indicative of the kind of character that had gotten the man elected.

The view in front of them, the river and the basin, still had not changed, but suddenly, as they watched, a section of city lights went out. It was the Vice-President who put it in words, "National Airport."

At first they saw only what could have been a flaw in the glass-like surface of the basin. It spread out like water poured on a floor, only in slow motion. The first flow did not appear to carry much water, but it did not stop, they could see the waterline continue to move. Soon the sidewalk around the Jefferson Memorial was touched with water from the basin.

To their far right another shimmering flow could be seen, just to the south of the Lincoln Memorial. This certainly was not as spectacular as the tsunami in the Pacific, but seeing these symbols of Americana slowly engulfed by the waters of global warming was dramatic.

The President turned his head to Andy, "Well, Andy, I guess this anoints you as a prophet."

"I am not sure what you mean."

"I had a long talk with the Speaker this afternoon. He told me about your harangue at the Conference Committee last month in which you predicted rowboat trips on the Mall."

"Yes, I thought about that meeting myself earlier this evening."

"I guess the new question is this; do you think all of this will convince them to pass the Greenhouse Gas Bill and the rest of the programs? Can we straighten things out before any more of this happens?"

"Unfortunately, Mr. President, I am not sure that really matters anymore. By the time this all ends," he waved his arm toward the water that now lapped at the edges of the Reflecting Pond, "it will have destroyed enough of our economy -- cities, factories, refineries, jobs, homes -- to easily reduce the greenhouse gases our economy produces several fold more than what my bill would have."

The President nodded silently.

Andy continued, "You remember how, when we first met in your office about this last Spring, I mentioned the Gaia hypothesis, the theory that all life on earth is a great self-regulated living organism?"

"Yes."

"Well, there are those who believe that this isn't simply a human catastrophe. If you ascribe to that theory, then this could just be the organism that is earth rebalancing the scales, settling into a new equilibrium of life. The symbology is that this is Gaia,

Mother Nature, the Goddess of the Earth, whatever you want to call her, restoring the equilibrium of the planet. Cleaning house after the excesses of her children and making sure that we children learn our lesson. Even if you don't tend toward the mythic, I guess it is pretty clear that we are being taught a lesson."

Then the President spoke, "While I firmly believe in the Gaia theory, my Catholic upbringing doesn't let me seriously consider any talk of Gaia, the earth goddess, as anything more than a symbol for the ecological theory. But, as I heard you speak of prophecy and gods, something occurred to me, right out of my memories of Sunday School class."

The President paused and continued, "It is ironic, a bit strange, and perhaps prophetic, that the government agency that tried to warn us of this Great Flood is called N.O.A.A.? And that we didn't listen to our Noah any better than the biblical people did."

Afterward by the Author

Santa Barbara, California [March, 1998]

This is a work of fiction and of the future. I can only pray that it is short on prophecy, but I fear it is not.

Although a work of fiction, this novel is founded exclusively on scientific and historical fact. With the exception of timing, in particular of the Ross Ice Shelf breakup, virtually every fact, estimation and prognosis upon which my story is based is well-founded in scholarly research and can be documented if the reader so wishes. For the layperson who wishes to find out more about the Gaia hypothesis, I can recommend as a good starting place GAIA, The Growth of an Idea by Lawrence E. Joseph, or, of course, any of James Lovelock's books. I must also recommend former Vice President Al Gore, Jr.'s work, Earth in the Balance.

This novel has, as any story must, fictional heroes and villains. Although a fictional upper hierarchy in the U.S. Commerce Department and the National Oceanographic and Atmospheric Agency (NOAA) are counted as villains in the early part of the story, in reality, the opposite is true. The help and information I received from the folks at NOAA, as well as from the U.S. Geological Survey, was invaluable in preparing this work. They have my sincere thanks.

This book is dedicated to James Lockwood, Lynn Margulis, Stephen Schneider, Al Gore, Jr. and the many others who have tried to warn us.

I also want to thank Toni Lorien for her editorial assistance.

Kevin E. Ready

About the Author

Kevin E. Ready

Kevin E. Ready studied Government and Politics at the University of Maryland, University College in Berlin, Germany and received his Juris Doctor degree with emphasis in Environmental Law from the University of Denver. He had four decades of experience as a US Navy officer, US Army officer and government attorney. He has twice been a major party candidate for US Congress. Kevin E. Ready lives in the Santa Barbara, California with his wife, Olga. He is the author and editor of several books.

—

Visit Kevin's website at http://www.KevinEReady.com
Kevin produces a blog on Politics and the Law at
http://www.LawfulPolitics.com and
http://www.Twitter.com/LawfulPolitics

Gaia Weeps

by

Kevin E. Ready

Published by

Saint Gaudens Press

Phoenix, Arizona — Santa Barbara, California

http://www.SaintGaudensPress.com

Saint Gaudens, Saint Gaudens Press
and the Winged Liberty colophon
are trademarks of Saint Gaudens Press
Copyright © 2020 Kevin E. Ready
Original Edition Copyright © 1998 Kevin E. Ready
All rights reserved.
eBook ISBN: 978-0-943039-00-8
Print edition ISBN: 978-0-943039-09-1
Printed in the United States of America

www.ingramcontent.com/pod-product-compliance
Lightning Source LLC
Chambersburg PA
CBHW021029210326
41598CB00016B/957